# CIVIL WAR WEATHER IN VIRGINIA

# CIVIL WAR WEATHER
# IN VIRGINIA

ROBERT K. KRICK

THE UNIVERSITY OF ALABAMA PRESS
*Tuscaloosa*

The University of Alabama Press
Tuscaloosa, Alabama 35487-0380
uapress.ua.edu

Hardcover edition published 2007.
Paperback edition published 2016.
eBook edition published 2007.

Inquiries about reproducing material from this work should be addressed
to the University of Alabama Press.

Typeface: AGaramond

Manufactured in the United States of America
Cover design: Todd Lape / Lape Designs

∞

The paper on which this book is printed meets the minimum requirements of
American National Standard for Information Science–Permanence of Paper for
Printed Library Materials, ANSI Z39.48-1984.

Paperback ISBN: 978-0-8173-5877-8
eBook ISBN: 978-0-8173-8048-9

A previous edition of this book has been catalogued by the Library of Congress as
follows:
Library of Congress Cataloging-in-Publication Data
Krick, Robert K.
Civil War weather in Virginia / Robert K. Krick.
p. cm.
Includes bibliographical references.
ISBN-13: 978-0-8173-1577-1 (cloth : alk. paper) ISBN-10: 0-8173-1577-2
1. Richmond (Va.)—History—Civil War, 1861–1865—Chronology. 2. Washington
(D. C.)—History—Civil War, 1861–1865—Chronology. 3. Richmond (Va.)—
History—Civil War, 1861–1865—Sources. 4. Washington (D. C.)—History—Civil
War, 1861–1865—Sources. 5. Richmond (Va.)—Climate—History—19th century.
6. Washington (D. C.)—Climate—History—19th century. 7. Meteorology—
Virginia—Richmond—History—19th century. 8. Meteorology—Washington
(D.C.)—History—19th century. 9. United States—History—Civil War, 1861–1865—
Chronology. 10. United States—History—Civil War, 1861–1865—Sources. I. Title.
F234.R557K75 2007
975.5'03—dc22
2006037503

# Contents

# Illustrations

# Acknowledgments

Several skilled reference librarians and historians helped me accumulate the material for this book. They include the redoubtable Michael P. Musick of the National Archives; Ham Dozier and Frances Pollard of the Virginia Historical Society; Jeffrey Edmunds, Barbara Pratt Willis, Ann Haley, Sue Willis, Jane Kosa, and Holly Schemmer of the Central Rappahannock Regional Library; Maggie Wilson and Ellen Ayers of the Smithsonian Institution; and Robert E. L. Krick, K. Sangria Bohannon, Eric J. Mink, Mac Wyckoff, C. Hollywood Ferguson, Francis A. O'Reilly, Keith E. Gibson, Tom Broadfoot, Gary W. Gallagher, and the late Lee A. Wallace Jr.

CIVIL WAR WEATHER IN VIRGINIA

# Introduction

"I beseech you," Douglas Southall Freeman said a few days before his death in an address to some avid battlefield tourers, "give us what we do not now have but long have needed, namely, a meteorological register of the War Between the States." The powerful impact of weather upon military operations suggested to Freeman the obvious importance of learning the details. "There are a great many days when we know nothing about the weather; we have no report of the temperature; we do not know whether the dust was rising or the mud was covering the men. . . . These are factors of the utmost importance and could be employed to inestimable advantage."[1]

More than half a century later, Freeman's plea remains unanswered. This book finally does what he suggested so long ago, although only for the Virginia theater of the war. The chances are high that the Virginia theater loomed, as always, at the forefront of Freeman's mind anyway, just as it does with me.

The diligence of an 1860s Presbyterian preacher in his sixties made this weather compilation possible. Most of the fundamental weather details of temperature and precipitation herein come from a ledger kept by the Reverend C. B. Mackee, who faithfully made readings in Georgetown, D. C., at 7 a.m., 2 p.m., and 9 p.m. each day. Mackee began the daily routine at Lewinsville in Fairfax County, Virginia, in June 1858. When he moved across the river into Georgetown sixteen months later, he carried the ledger with him and kept on recording his observations.[2]

In June 1860 the census found "C. B. MacKee [sic]," age sixty-five, living in Georgetown's Third Ward. Mackee[3] gave his occupation as "O.S.P. Minister." Instructions for the 1860 census recorders admonished them to abbreviate religious denominations. Fortunately, one of only three examples supplied to illustrate that system was "O.S.P.," to mean Old School Presbyterian. The Reverend Mackee gave his place of birth as Pennsylvania, and the value of his total personal estate as one hundred dollars, a very scant estate indeed, even in that era. Mrs. Hanna Mackee, age forty, had been born in Iceland. The five children, three of them girls, ranged downward from a child age eleven, born in Maryland, to an infant age one, born in Virginia.[4]

Mackee's ledgers survive today thanks to a concerted effort by the Weather Bureau, mostly in the 1950s, to secure surviving copies of early records for inclusion in a national meteorological archive. The front matter on the microfilmed roll that includes Georgetown offers this introductory explanation: "Weather Diaries and Journals loaned by the National Archives to the Weather Bureau for microfilming. Most of the sources were collected by the Smithsonian Institution or the Army Signal Service prior to the establishment of the Weather Bureau in 1891."[5]

All but two of the fifty-seven tables in this book, which provide basic weather data for the months from October 1860 through June 1865, report exclusively the temperature and precipitation details collected by Rev. Mackee in Washington. A broken thermometer knocked the Georgetown observer out of business for three weeks in February 1861; a Richmond station fills that gap, as is explained in some detail at the appropriate point below. The complete absence of Georgetown's weather-ledger sheet for August 1863 creates the only other anomaly, filled in that instance by reports from just across the river at Alexandria, Virginia. The details of that necessary adjustment appear in the August 1863 entry.

The importance of providing a uniform set of data made it easy to dodge the temptation to compile the tables for various months from various weather stations. None of those other stations operated for more than one-fourth of this period, most for a shorter span than that, while Georgetown's reports survive for 96 percent of the 1,732-day period from October 1, 1860, to June 30, 1865—a total of more than five thousand individual readings. Interesting material from Richmond, Cumberland, Lexington, and elsewhere shows up in this book on the monthly pages that face the tables. Even in instances when an individual Georgetown ledger entry is missing or illegible, and another station can fill the gap, that information is supplied in the facing prose page, not inserted directly into the ledger (cf. April 20–21, 1864). Excepting only February 1861 and August 1863, temperature or comments in the tables herein always comes from Georgetown.

Throughout this book, the station in Georgetown, Washington, D.C., is referred to randomly as either Georgetown or Washington. The origins of the Georgetown manuscript will not be cited hereafter, although every use of any other weather manuscript will result in a note citation. All geographical locations are in Virginia if not otherwise specified.

The steady use of temperatures from the Georgetown observer surely will prompt some reviewer to grumble that, because of the few hundred yards from Georgetown to the Virginia border, the title of this book, *Civil War Weather in Virginia*, does not represent the content properly. In fact, were the title page being framed in the wordy style of the Victorian era, it might well and accurately read: *Civil War Weather in Washington, D.C., and in the Virginia Theater of War, Encompassing Virginia and Maryland and Pennsylvania, and Including Weather Affecting Some Military Operations in What Became West Virginia Halfway through the War.*

All fifty-seven monthly tables provide times of sunrise and sunset in Richmond (not Washington), with notes that cite the almanac sources for that information reported at the first month of each year. The dates of solstice and equinox also come from the almanacs.

Hanna Mackee, or perhaps even the older children, may have helped Rev. Mackee take or record the readings. The handwriting usually does remain constant, and consistently ragged, through much of the period. A note on May 1, 1865, reads: "Pa left home for New York." Presumably the oldest daughter, by then sixteen years old, made the note; or maybe Hanna referred to her two-decades-older husband as "Pa."

Entries on the Mackee weather sheets usually could be deciphered without difficulty. The biggest problem with reading the handwriting came in the case of the numerals o and 6. An intermittent tendency toward sloppy conclusion of a o at its upper right corner, combined with a 6 that crawls high with its final stroke, provoked some discomfort during the transcriptions. For about half of the years under study, the Mackees added up their three daily readings, divided by three, and recorded the average. That convenient amenity made possible arithmetic double-checking of some uncertain temperatures. The occasional entry defaced or written over repeatedly, and thus rendered illegible, also could be established firmly by interpolation from the daily average. Unfortunately, in the later years the Mackees apparently grew weary of doing the math, and stopped supplying the average.

Sometimes the Georgetown temperatures appear on the original sheets with half degrees. In most of those instances I have rounded to the nearest even number. At times when partial numbers have seemed worth using (with rainfall, for instance, or monthly averages), I have converted most fractions into decimals.

Comments about rain or other events appear in the Mackee manuscripts on a second page, opposite a long array of figures not included in my tables (figures that report wind, clouds, etc.). For some months, it is difficult—occasionally even impossible—to be sure of the alignment between the pages. In a few instances, fortunately not frequent, the offset made it impossible to use the weather events reported on the second page. Accordingly, someone using this reference book should *not* assume that the absence of comment about rain or storm means that there was none. What you do see should be accurate; the absence of comment does not necessarily mean an absence of things worthy of comment. When the Mackees copied dates onto page two, next to their comments (in August and September 1864, for example), the remarks achieve complete reliability, and comments facing the table usually report that. On many other months, the comments occurred so frequently that they provide plenty of match points of their own (November and December 1861, for example), which also ensures accuracy.

Reporting of rainfall totals in most instances appears as a decimal. When did that rain fall? We cannot be absolutely certain, but obviously it usually would not be measured until the precipitation ended. On some monthly ledgers Mackee added a column for rain, headed "24 hours." Clearly his intention was to report the rainfall at the *end* of twenty-four hours, or at least the evident end of a storm. Some of the precipitation reports in the ledgers make clear the timing of the rainfall, typically over the night just past. The May 4, 1861, reading obviously means that .83 inches of rain fell "all last night," and not primarily on May 4. October 18, 1861, conveys a similar message. So do many other days. Some readings necessarily will be obscure, based on the original entries. The tables show what the ledgers reported. Fortunately, most of the information falls into understandable patterns.

The tables that supply the hard data in this book come almost entirely from the single Georgetown station described above, except during the very brief lacunae in its records. I have used other regional records to cover those short gaps. Entries also survive from other recording stations in the region, notably in the holdings of both the National Archives and (to a limited degree) the Smithsonian Institution. My purpose has been to supply information to students of the Civil War about weather during that cataclysm, but this work makes no pretensions to being, and cannot be called, a serious weather book. Were its purpose to declaim about weather for scientific reasons, using a single recording station would be foolish.

Students eager to find weather records in diverse detail for this period, or any other, will find a bonanza in Record Group 27 at the National Archives. Lewis J. Darter Jr. compiled a thorough finding aid for that record group in March 1942: *List of Climatological Records in the National Archives.* Darter traces federal record keeping of weather data from its roots in 1814, in the office of the surgeon general of all places. He divides existing records into a useful grouping by state, including ten locations in the District of Columbia (only two of them—one being the Georgetown site—spanning the Civil War years). Of the 112 weather stations in Virginia for which Darter found National Archives records, only three include any substantive readings for the Civil War years. Many only began recording after 1865. Those with prewar readings almost invariably succumbed to the same dislocation that the war wreaked on everything else in its insensate path.[6]

The Smithsonian Institution played a pivotal role in collecting weather data before the establishment of a national weather service. Anyone attempting to be thorough about weather data before 1873 will find useful material accumulated by the Smithsonian Institution—but not now held by that institution. Most of those Smithsonian records have made their way into federal files at the National Archives. Darter's guide describes an imposing eighty shelf-feet of Smithsonian weather records that were subsumed into Record Group 27.[7]

The narrative section for each month in this book, appearing after each monthly table, supplies weather commentary from soldiers in the field and from civilian letters and diaries. Those excerpts of course do not pretend to be comprehensive, offering as they do only scattered examples from the vast range available. In the case of primary accounts by soldiers, the sources in most cases are somewhat obscure, under the premise that some readers will be less familiar with them than they would be with more widely available accounts. Users of this book will discover the narrative pages to be fragmentary samplings—more than random but nothing remotely like definitive, although designed to be useful.

Quotes supplied in the narrative pages are strictly verbatim, excepting only the occasional change of capitalization to conform to sentence structure. Ellipses indicate any lacunae within direct quotes, but they have not been used at the beginning or end of quotes that truncate larger sentences in the original sources.

Diarists and correspondents almost inevitably described bad weather and ignored good weather, unless it was particularly spectacular or unexpected. "How prone we are to note only the *stormy* days," a Virginian wrote in his diary in 1862.[8] That unmistakable tendency affects the nature of available evidence, but it is the anomalies that interest us primarily in any event.

The impact of weather on military operations makes reliable weather statistics an important historical tool. The information printed here for the first time should be useful to students and historians at work on any event in or near Virginia during the fifty-seven months that it covers. Weather has at least some effect on everything soldiers do. Occasionally it levies more impact than do strategy or tactics or materiel. The eponymous viscous gumbo of Ambrose E. Burnside's "Mud March" ruled that week in January 1863. The Romney Campaign in January 1862 suffered negligibly from Yankee reactions, but dreadfully from wintry blasts. Less obviously, Stonewall Jackson's daring flank march at Chancellorsville could hardly have maintained its vital secrecy had it stirred up a betraying column of dust—but the woods paths he used for his clandestine march had been conveniently dampened by showers on April 28–29, 1863.

In a contrary vein, the hard data reproduced here demonstrate that traditional lore about weather during some Civil War episodes has been exaggerated, even fabricated. The incessantly repeated fiction about wounded freezing to death in the aftermath of the Battle of Fredericksburg apparently began in the imagination of someone who supposed that mid-December must have been frigid. In fact, December 13–15, 1862, warmed up enough to warrant the "Indian summer" label that some contemporaries applied to the period.

This work makes no attempt to evaluate the unmistakable effect of weather on Civil War activities in Virginia. That worthy and interesting topic deserves further elaboration, but such an examination is not among the goals of this book. The contents herein should be invaluable for such an endeavor—but this book is designed for reference purposes, not analytical ones. And estimable recent work that is devoted to considering weather's impact on war is Harold A. Winters et al., *Battling the Elements: Weather and Terrain in the Conduct of War* (Baltimore: Johns Hopkins University Press, 1998). A few sections of modest scope in *Battling the Elements* address Civil War operations, but the book's broad latitudes extend from Kublai Khan to Dien Bien Phu. Not surprisingly, the cover illustration depicts Bonaparte riding through Russian snows in what probably is the best-known triumph of the elements over military affairs.

# Chapter 1

# 1860

October 1860

A Washington diarist recorded her thoughts on the capital city's weather on several October days. October 3: "Delightful day, atmosphere very bracing." 4th: "Clear and cold." 5th: "Raining slightly." 24th: "Heavy frost last night, fear some of our flowers were injured." 26th: "Delightfully mild today." 31st: "Raining hard all day."[1]

A weather station reporting from Virginia's capital city provides steady and reliable information through February 1862, when it ceased operation. Charles J. Meriwether kept that ledger at his home, "Westwood," on the Brook Turnpike just north of Richmond. Meriwether made his entries on printed sheets supplied by the Smithsonian Institution, with the printed declaration, "Adopted by the Commissioner of Patents for his Agricultural Report." Meriwether took temperature readings thrice daily, at the same times as the Georgetown recorder: 7 a.m., 2 p.m., and 9 p.m. The average for October at those times was 53.8, 67.4, and 59.1. During this extremely wet month, Richmond reported rain on October 1, 4, 5, 8, 11, 14, 19, 20, 30, and 31. The heaviest storm dropped 1.5 inches on the 19th and 20th.[2]

In Fredericksburg, after "beautiful weather" on October 18, a "violent storm" raged on October 19 from 4 p.m. to midnight. A local newspaper described "a cruel storm of rain and wind."[3]

## October 1860

| Date | Day of Week | Rich-mond Sunrise | Rich-mond Sunset | D.C. Temperature | | | Other |
|---|---|---|---|---|---|---|---|
| | | | | 7 a.m. | 2 p.m. | 9 p.m. | |
| 1 | Mon. | 6:10 | 5:50 | 52 | 62 | 59 | .75 rain, fair at 2 |
| 2 | Tues. | 6:11 | 5:49 | 58 | 82 | 73 | air balmy warm |
| 3 | Wed. | 6:12 | 5:48 | 64 | 71 | 66 | |
| 4 | Thurs. | 6:13 | 5:47 | 62 | 67 | 63 | drizzled off and on all day |
| 5 | Fri. | 6:14 | 5:46 | 62 | 77 | 69 | heavy fog til 6 |
| 6 | Sat. | 6:16 | 5:44 | 57 | 66 | 52 | |
| 7 | Sun. | 6:17 | 5:43 | 47 | 63 | 61 | at 8 p.m. beginning rain |
| 8 | Mon. | 6:18 | 5:42 | 62 | 72 | 62 | rained last night .46 |
| 9 | Tues. | 6:19 | 5:41 | 48 | 60 | 50 | |
| 10 | Wed. | 6:21 | 5:39 | 47 | 68 | 59 | |
| 11 | Thurs. | 6:22 | 5:38 | 56 | 68 | 58 | raining at 9 p.m. |
| 12 | Fri. | 6:23 | 5:37 | 48 | 58 | 46 | |
| 13 | Sat. | 6:24 | 5:36 | 38 | 58 | 50 | |
| 14 | Sun. | 6:26 | 5:34 | 46 | 45 | 43 | drizzled all day .80 |
| 15 | Mon. | 6:27 | 5:33 | 38 | 57 | 45 | |
| 16 | Tues. | 6:28 | 5:32 | 37 | 54 | 47 | |
| 17 | Wed. | 6:29 | 5:31 | 42 | 56 | 52 | |
| 18 | Thurs. | 6:30 | 5:30 | 54 | 60 | 58 | showered at 3 |
| 19 | Fri. | 6:31 | 5:29 | 52 | 56 | 50 | damp, rainlike |
| 20 | Sat. | 6:32 | 5:28 | 54 | 60 | 56 | rained last night 1.62 |
| 21 | Sun. | 6:33 | 5:27 | 58 | 68 | 56 | heavy fog, sunshine at 10 |
| 22 | Mon. | 6:34 | 5:26 | 47 | 67 | 56 | rained last night .15 |
| 23 | Tues. | 6:36 | 5:24 | 53 | 64 | 58 | rain at 7 |
| 24 | Wed. | 6:37 | 5:23 | 57 | 66 | 54 | |
| 25 | Thurs. | 6:38 | 5:22 | 5- | 66 | 57 | very damp |
| 26 | Fri. | 6:39 | 5:21 | 50 | 74 | 60 | very heavy dews |
| 27 | Sat. | 6:40 | 5:20 | 50 | 64 | 54 | |
| 28 | Sun. | 6:41 | 5:19 | 52 | 63 | 60 | rainlike all day |
| 29 | Mon. | 6:43 | 5:17 | 59 | 65 | 66 | rainlike |
| 30 | Tues. | 6:44 | 5:16 | 64 | 70 | 65 | rained last night .35 |
| 31 | Wed. | 6:45 | 5:15 | 64 | 68 | 67 | rained last night .50 |

November 1860

A Fredericksburg newspaper recognized the inevitable impact of weather on voting habits, and in a November 2 column encouraged its readers not to be daunted should election day prove unpleasant: "You should go to the polls . . . no matter what the weather."[4]

In King & Queen County, November 5 was "quite cool," with "a large frost" in the morning. On the 12th, two boys out hunting there "got as wet as rats."[5]

Weather did not interfere with voting in most Virginia locales. A few high clouds scurried across election-day skies on Tuesday the 6th in Smithfield, Isle of Wight County. A weather station in that Tidewater community recorded temperatures on election day of 45, 60, and 45 at 7 a.m., 2 p.m., and 9 p.m.[6]

On the 14th, a Washington woman wrote, "things look very ominous politically," and a "beautiful morning" became "cloudy at noon."[7]

In Charlottesville, November 25 dawned "both wet & cold." A pious woman concluded with dismay that the day was "so very cold" that going to church would be impracticable.[8]

Charles Meriwether atypically filled in his ledger sheet only sparsely during this month at Richmond. He recorded a heavy rain that began during the night on November 27 and ended at noon on the 28th, dropping 1.375 inches on the city.[9]

The Georgetown station reported nine-tenths (0.9) of an inch of rain on November 9. The tabulation of rainfall for the month, however, reached only 2.28 inches. For that calculation to make sense, the November 9 total would have to have been nine-hundredths (.09) of an inch. Since the comment for that day includes "very hard" rain, the chances are that .90 is correct, and the error lies in the end-of-month mathematics, which then should total 3.09.

November 1860

| Date | Day of Week | Rich-mond Sunrise | Rich-mond Sunset | D.C. Temperature | | | Other |
|---|---|---|---|---|---|---|---|
| | | | | 7 a.m. | 2 p.m. | 9 p.m. | |
| 1 | Thurs. | 6:46 | 5:14 | 66 | 75 | 68 | .15 rain, cleared up in p.m. |
| 2 | Fri. | 6:47 | 5:13 | 65 | 72 | 58 | showery, occasional sun |
| 3 | Sat. | 6:48 | 5:12 | 65 | 65 | 58 | 1.34 rain, cleared at dusk |
| 4 | Sun. | 6:49 | 5:11 | 48 | 58 | 46 | a few clouds |
| 5 | Mon. | 6:50 | 5:10 | 42 | 62 | 54 | heavy white frost, pleasant |
| 6 | Tues. | 6:51 | 5:09 | 45 | 62 | 41 | clear day |
| 7 | Wed. | 6:52 | 5:08 | 40 | 50 | 40 | |
| 8 | Thurs. | 6:53 | 5:07 | 35 | 52 | 44 | heavy frost |
| 9 | Fri. | 6:54 | 5:06 | 56 | 56 | 50 | .90 rain, very hard at noon |
| 10 | Sat. | 6:55 | 5:05 | 46 | 46 | 58 | drizzling from noon |
| 11 | Sun. | 6:56 | 5:04 | 50 | 54 | 48 | gloomy, windy |
| 12 | Mon. | 6:57 | 5:03 | 43 | 68 | 49 | damp, gloomy day |
| 13 | Tues. | 6:58 | 5:02 | 43 | 58 | 52 | remarkably clear day |
| 14 | Wed. | 6:59 | 5:01 | 45 | 62 | 48 | a clear, calm day |
| 15 | Thurs. | 7:00 | 5:00 | 39 | 62 | 44 | |
| 16 | Fri. | 7:01 | 4:59 | 37 | — | 50 | |
| 17 | Sat. | 7:01 | 4:59 | 48 | 54 | — | very heavy fog, .35 rain |
| 18 | Sun. | 7:02 | 4:58 | 50 | 56 | — | |
| 19 | Mon. | 7:03 | 4:57 | 46 | 60 | — | |
| 20 | Tues. | 7:04 | 4:56 | 44 | 48 | 43 | |
| 21 | Wed. | 7:05 | 4:55 | 37 | 42 | 36 | gloomy |
| 22 | Thurs. | 7:06 | 4:54 | 31 | 44 | 40 | |
| 23 | Fri. | 7:07 | 4:53 | 42 | 42 | 42 | .35 rain |
| 24 | Sat. | 7:07 | 4:53 | 30 | 28 | 21 | windy night, slight snow |
| 25 | Sun. | 7:08 | 4:52 | 18 | 31 | 26 | |
| 26 | Mon. | 7:09 | 4:51 | 22 | 42 | 40 | damp & snowlike |
| 27 | Tues. | 7:09 | 4:51 | 41 | 52 | 50 | drizzly all day |
| 28 | Wed. | 7:10 | 4:50 | 45 | 45 | 33 | drizzly, cleared at 3 |
| 29 | Thurs. | 7:11 | 4:49 | 30 | 45 | 48 | lurid sky & cold a.m. |
| 30 | Fri. | 7:11 | 4:49 | 41 | 46 | 40 | rained very hard midday |

## December 1860

"Wind blowing, very cold" in Washington on December 2. The next day the weather had "moderated," but on the 4th "the first snow of the season" fell. Snow covered the ground in the nation's capital on December 14, and more snow fell on the 18th and 24th; by the last date, the nation centered on Washington had shrunk by one state, at least in the view of South Carolinians. Christmas brought "very cold" weather, and on the 28th, "snow very deep."[10]

The season's first snow also fell on Fredericksburg the night of December 3–4, but later on the 4th the "sun [was] shining brightly."[11]

A youngster in King and Queen County noted in his diary on December 7 that it had been "quite cold all day, snowed a little this evening." Rain fell there on the night of December 21 and into the morning of the 22nd. On Christmas Day, temperatures were low enough to make skating on a farm pond practicable.[12]

At daybreak on December 15, a Charlottesville woman "awoke . . . to see the ground covered with a beautiful light covering of snow. . . . Tis the first time the snow has lain on the grown for two years."[13]

In Richmond, the month-long temperature readings averaged 31.9, 42.0, and 35.4 degrees at 7 a.m., 2 p.m., and 9 p.m. As South Carolina seceded on December 20, the thermometer in Richmond read 40, 52, and 44 degrees at those three hours. Richmond's first snow, as might be expected, fell on the same day as Washington's, one inch on December 4. The Richmond station recorded two other snows during the month, 1.75 inches on December 19, and six inches on the 31st.[14]

Two Decembers hence, mighty armies gathered near Fredericksburg would see an aurora borealis under memorable circumstances in the aftermath of a great battle. The weather keeper in Fredericksburg noted an aurora on this December 10; the coldest morning yet on the 14th; "heavy snow" on the 15th; and a "general thaw" on the 20th.[15]

A blizzard blanketed the upper reaches of Virginia's Shenandoah Valley with 17 inches of snow on December 29 and 30.[16]

## December 1860

| Date | Day of Week | Rich-mond Sunrise | Rich-mond Sunset | D.C. Temperature | | | Other |
|------|------|------|------|------|------|------|------|
| | | | | 7 a.m. | 2 p.m. | 9 p.m. | |
| 1 | Sat. | 7:12 | 4:48 | 34 | 34 | 30 | snowlike, black frost |
| 2 | Sun. | 7:12 | 4:48 | 28 | 37 | 33 | high NW winds |
| 3 | Mon. | 7:13 | 4:47 | 31 | 40 | 34 | |
| 4 | Tues. | 7:13 | 4:47 | 34 | 38 | 30 | snow last night, sun today |
| 5 | Wed. | 7:14 | 4:46 | 29 | 42 | 35 | melting fast |
| 6 | Thurs. | 7:14 | 4:46 | 34 | 41 | 37 | |
| 7 | Fri. | 7:15 | 4:45 | 32 | 46 | 40 | |
| 8 | Sat. | 7:15 | 4:45 | 38 | 39 | 37 | .25 rain last night & today |
| 9 | Sun. | 7:15 | 4:45 | 37 | 42 | 37 | blue & golden |
| 10 | Mon. | 7:16 | 4:44 | 37 | 43 | 46 | .41 rain |
| 11 | Tues. | 7:16 | 4:44 | 42 | 45 | 33 | sun rose golden but occ. snow |
| 12 | Wed. | 7:16 | 4:44 | 29 | 46 | 40 | heavy frost |
| 13 | Thurs. | 7:17 | 4:43 | 34 | 47 | 34 | a few snowflakes |
| 14 | Fri. | 7:17 | 4:43 | 19 | 27 | 22 | cold day, Rock Creek frozen over |
| 15 | Sat. | 7:17 | 4:43 | 18 | 22 | 20 | snowed all day, 2.5 |
| 16 | Sun. | 7:17 | 4:43 | 14 | 30 | 37 | calm cold |
| 17 | Mon. | 7:18 | 4:42 | 20 | 41 | 30 | |
| 18 | Tues. | 7:18 | 4:42 | 26 | 41 | 34 | |
| 19 | Wed. | 7:18 | 4:42 | 32 | 38 | 40 | drizzling |
| 20 | Thurs. | 7:18 | 4:42 | 41 | 47 | 45 | heavy rain, 3–4 a.m. |
| 21 | Fri. | 7:18 | 4:42 | 34 | 48 | 42 | Winter Solstice |
| 22 | Sat. | 7:18 | 4:42 | 42 | 45 | 32 | |
| 23 | Sun. | 7:18 | 4:42 | 30 | 36 | 32 | |
| 24 | Mon. | 7:18 | 4:42 | 25 | 30 | 28 | |
| 25 | Tues. | 7:18 | 4:42 | 28 | 34 | 32 | spit snow, .5 |
| 26 | Wed. | 7:18 | 4:42 | 30 | 37 | 32 | |
| 27 | Thurs. | 7:17 | 4:43 | 30 | 33 | 32 | |
| 28 | Fri. | 7:17 | 4:43 | 32 | 34 | 32 | snowlike |
| 29 | Sat. | 7:17 | 4:43 | 30 | 38 | 34 | snowlike |
| 30 | Sun. | 7:17 | 4:43 | 35 | 41 | 38 | drizzled all day, .57 |
| 31 | Mon. | 7:17 | 4:43 | 30 | 35 | 22 | rain & snow last night, 1.5 |

# Chapter 2

# 1861

## January 1861

The New Year dawned in Washington clear, but "growing colder every minute." Rain fell all night on the 2nd and into the morning of the 3rd. On January 8, which was "snowing and raining," a woman in D.C. grumbled: "We have had a great deal of damp and gloomy weather this winter." The diarist also mentioned snow on January 11, 14, and 25.[1]

Average temperatures in Richmond in January paralleled the December readings remarkably closely: 31.9, 43.2, and 37.4 degrees at 7 a.m., 2 p.m., and 9 p.m. New Year's opened in Virginia's capital city at 10 degrees at 7 a.m., reached 35 at 2 p.m., and dropped to 27 at 9 p.m. Snow fell on January 14–16, 23, and 26 (3.5 inches on the last date), and rain on January 9 and 24 (1.25 inches overnight).[2]

On January 11 in Richmond, "the atmosphere . . . set everyone to shivering and the general opinion was that a snow storm was somewhere about."[3]

Predictably, far colder temperatures prevailed in the higher elevations of the upper Shenandoah Valley. At 7 a.m. on both January 1 and 2 the thermometer at the Virginia Military Institute (VMI) in Lexington read 5 degrees. As Professor Thomas J. Jackson of the VMI strode between his home and his section room that month, presumably after his customary cold bath, temperatures at

7 a.m., 2 p.m., and 9 p.m. averaged 27.5, 36.1, and 31.2 degrees; the highest reading was 48.8 on the 16th. After the deep snow of December 29–30, Lexington escaped with only mixed snow and sleet on January 14 and 15 and snow turning to rain on the 24th. The town's rainfall for January totaled 3.9 inches.[4]

A correspondent writing from Bowling Green referred to the weather there on January 13 as "inclement."[5]

In Virginia's western mountains, a Greenbrier County correspondent noted mid-month that "the weather for some days has been clear, cold and calm. The snow is disappearing." Neither the Greenbrier River, however, nor any of its extensive tributaries had yet been frozen over during the winter.[6]

## January 1861

| Date | Day of Week | Rich-mond Sunrise | Rich-mond Sunset | D.C. Temperature | | | Other |
|------|------|------|------|------|------|------|------|
| | | | | 7 a.m. | 2 p.m. | 9 p.m. | |
| 1 | Tues. | 7:16 | 4:44 | 14 | 34 | 28 | occasional clouds |
| 2 | Wed. | 7:16 | 4:44 | 20 | 36 | 36 | heavy frost, foggy |
| 3 | Thurs. | 7:15 | 4:45 | 36 | 40 | 28 | rained all night .92 |
| 4 | Fri. | 7:15 | 4:45 | 32 | 40 | 36 | rained thru night |
| 5 | Sat. | 7:15 | 4:45 | 26 | 40 | 32 | calm day |
| 6 | Sun. | 7:14 | 4:46 | 26 | 41 | 36 | smoky horizon |
| 7 | Mon. | 7:14 | 4:46 | 39 | 50 | 42 | raining slightly |
| 8 | Tues. | 7:13 | 4:47 | 40 | 48 | 38 | |
| 9 | Wed. | 7:12 | 4:48 | 38 | 38 | 30 | snowed and rained from 9 to 12 |
| 10 | Thurs. | 7:12 | 4:48 | 32 | 42 | 32 | heavy frost |
| 11 | Fri. | 7:11 | 4:49 | 23 | 30 | 30 | froze hard last night |
| 12 | Sat. | 7:11 | 4:49 | 32 | 40 | 24 | |
| 13 | Sun. | 7:10 | 4:50 | 12 | 31 | 17 | |
| 14 | Mon. | 7:09 | 4:51 | 6 | 31 | 30 | snowed slightly, sleeting |
| 15 | Tues. | 7:09 | 4:51 | 32 | 35 | 34 | drizzled all day |
| 16 | Wed. | 7:08 | 4:52 | 36 | 40 | 40 | rained heavily all night |
| 17 | Thurs. | 7:07 | 4:53 | 37 | 44 | 34 | sleeted and rained all day |
| 18 | Fri. | 7:06 | 4:54 | 32 | 36 | 35 | |
| 19 | Sat. | 7:06 | 4:54 | 34 | 45 | 44 | severe frost, p.m. springlike |
| 20 | Sun. | 7:05 | 4:55 | 36 | 36 | 32 | |
| 21 | Mon. | 7:04 | 4:56 | 32 | 44 | 32 | |
| 22 | Tues. | 7:03 | 4:57 | 26 | 38 | 28 | |
| 23 | Wed. | 7:03 | 4:57 | 24 | 32 | 32 | |
| 24 | Thurs. | 7:02 | 4:58 | 32 | 34 | 40 | rained all day |
| 25 | Fri. | 7:01 | 4:59 | 35 | 42 | 38 | |
| 26 | Sat. | 7:00 | 5:00 | 32 | 40 | 28 | snowed 7 inches |
| 27 | Sun. | 6:59 | 5:01 | 26 | 38 | 28 | calm & clear |
| 28 | Mon. | 6:58 | 5:02 | 25 | 37 | 25 | |
| 29 | Tues. | 6:57 | 5:03 | 32 | 48 | 44 | |
| 30 | Wed. | 6:56 | 5:04 | 40 | 31 | 28 | |
| 31 | Thurs. | 6:55 | 5:05 | 18 | 36 | 24 | |

February 1861

February opened in Washington "cold and raw." After cold rain all night on the 1st, "fog so thick that I cannot see . . . across the street" blanketed the city on the morning of the 2nd. Snow on February 4 and 10 yielded to "our first spring-like day" on the 14th. On the morning of the 24th, with president-elect Lincoln newly arrived in town, a front brought in "high winds, threatening rain."[7]

Sometime on February 7, the Georgetown weather observer broke his thermometer. For the next three weeks he took no readings at all. The comments did continue in the Georgetown ledger, as the final column in the February table shows. The table's temperature readings, however, from 2 p.m. on February 7 through 9 p.m. for February 27, come from the Richmond weather report. To avoid any confusion, the table includes "R" after each temperature reading from Richmond. The Richmond temperatures entered in this table are the only entries in the tables that are not from Washington, except for the month of August 1863. (All months use Richmond sunrise and sunset times.)[8]

Near the crest of the Blue Ridge at Paris and Ashby's Gap, February 2 seemed "bitter cold."[9]

Richmond's February afforded far warmer temperatures than the preceding two months, averaging 53.1 degrees at 2 p.m. Eight days in the 50s, eight more in the 60s, and 71 degrees on February 28 marked a very mild month. The station measured precipitation on six days, and its lowest recorded 2 p.m. reading was 25 degrees on February 8.[10]

Downriver from Richmond in Tidewater Virginia, temperatures as usual registered a bit higher. At Smithfield, the averages at 7 a.m., 2 p.m., and 9 p.m. were 38.3, 55.6, and 42.7.[11]

Weather hampered voters in the elections in Clarkesville, Mecklenburg County, as they chose delegates to the Richmond convention. "The Roanoke [River] is higher than it has been since 1850," a local man wrote on February 4.[12]

Rain fell steadily during February on Lexington; twelve days of rain yielded 3.71 inches. The lowest recorded temperature of the month there was 18.1 degrees early on February 8, and the highest was 65.8 degrees on the 28th. Temperatures taken at 7 a.m., 2 p.m., and 9 p.m. averaged 32.4, 44.6, and 38.4 degrees.[13]

A Fredericksburg journalist writing on March 1 reported cheerfully, if unscientifically, that "the winter . . . closed its existence beautifully yesterday."[14]

## February 1861

| Date | Day of Week | Rich-mond Sunrise | Rich-mond Sunset | D.C./Richmond Temp. 7 a.m. | 2 p.m. | 9 p.m. | Other (DC only) |
|---|---|---|---|---|---|---|---|
| 1 | Fri. | 6:54 | 5:06 | 25 | 36 | 47 | rained slightly |
| 2 | Sat. | 6:53 | 5:07 | 46 | 50 | 50 | heavy fog all day |
| 3 | Sun. | 6:52 | 5:08 | 40 | 40 | 37 | snow all gone |
| 4 | Mon. | 6:51 | 5:09 | 34 | 44 | 34 | snowed all night 1.5 |
| 5 | Tues. | 6:50 | 5:10 | 28 | 38 | 34 | |
| 6 | Wed. | 6:49 | 5:11 | 32 | 48 | 32 | slightly foggy all day |
| 7 | Thurs. | 6:48 | 5:12 | 40 | 58R | 34R | thermometer broken, very high wind |
| 8 | Fri. | 6:47 | 5:13 | 12R | 25R | 22R | very cold, water froze in the kitchen |
| 9 | Sat. | 6:46 | 5:14 | 22R | 46R | 40R | calm |
| 10 | Sun. | 6:45 | 5:15 | 39R | 66R | — | damp & rainlike |
| 11 | Mon. | 6:44 | 5:16 | 48R | 65R | — | damp, gloomy day |
| 12 | Tues. | 6:43 | 5:17 | 57R | 64R | — | rained heavily all last night |
| 13 | Wed. | 6:42 | 5:18 | 41R | 63R | — | calm day |
| 14 | Thurs. | 6:41 | 5:19 | 39R | 63R | — | occ. gusts of wind |
| 15 | Fri. | 6:40 | 5:20 | 56R | 57R | — | showers all day .15 |
| 16 | Sat. | 6:39 | 5:21 | 36R | 57R | 50R | |
| 17 | Sun. | 6:37 | 5:23 | 36R | 43R | 38R | began to spit snow 1 p.m. |
| 18 | Mon. | 6:36 | 5:24 | 31R | 42R | 34R | |
| 19 | Tues. | 6:35 | 5:25 | 27R | 43R | 38R | damp & cold all day |
| 20 | Wed. | 6:34 | 5:26 | 40R | 52R | 42R | snowed & rained last night |
| 21 | Thurs. | 6:32 | 5:28 | 35R | 47R | 40R | |
| 22 | Fri. | 6:31 | 5:29 | 36R | 54R | 40R | |
| 23 | Sat. | 6:30 | 5:30 | 50R | 62R | 59R | sprinkling |
| 24 | Sun. | 6:29 | 5:31 | 52R | 52R | 35R | windy last night |
| 25 | Mon. | 6:28 | 5:32 | 26R | 45R | 38R | |
| 26 | Tues. | 6:27 | 5:33 | 39R | 64R | — | |
| 27 | Wed. | 6:26 | 5:34 | 38R | 67R | — | |
| 28 | Thurs. | 6:25 | 5:35 | 42 | 70 | 61 | |

March 1861

Mrs. E. L. Lomax's Washington diary provides intermittent weather updates. 6th: "Still windy. . . . much colder." 11th: "Cloudy." 14th: "Snowing and thawing." 18th: "Very cold and cloudy." 19th: "Bright sun. . . . Ground well covered with snow." Easter Sunday, the 31st: "A bright day, full of sunshine."[15]

One hundred miles south of Washington, Richmond's last full pre-Confederate month brought seasonable temperatures, averaging 42, 57, and 48.5 degrees at 7 a.m., 2 p.m., and 9 p.m. Rain fell there on March 9, 14, 17, 18, 19, 20, and 27.[16] Averages at the same hours in Lexington, where Thomas J. Jackson (soon to earn the nom de guerre "Stonewall") was spending his last month as a professor, averaged 37.9, 49.6, and 43.7.[17]

Midway between the capitals, a snowstorm hit Fredericksburg at 5 p.m. on March 14, and a heavy thunderstorm with hail struck on the 27th. The storm cleared off by the morning of the 28th, which became "cold and bracing." On March 8 and 9 the Fredericksburg weatherman observed an aurora borealis, precursor of the famous event seen from the same city in December 1862. He described the spectacle on the first night as "quite bright" between 8 and 10 p.m.: "seemed to be a brilliant fog without arches or beams."[18]

In Tidewater Virginia, temperatures at Smithfield at 7 a.m., 2 p.m., and 9 p.m. averaged 41.9, 57.5, and 46 degrees during March.[19]

## March 1861

| Date | Day of Week | Rich-mond Sunrise | Rich-mond Sunset | D.C. Temperature | | | Other |
|---|---|---|---|---|---|---|---|
| | | | | 7 a.m. | 2 p.m. | 9 p.m. | |
| 1 | Fri. | 6:23 | 5:37 | 51 | 73 | 69 | |
| 2 | Sat. | 6:22 | 5:38 | 67 | 81 | 66 | |
| 3 | Sun. | 6:21 | 5:39 | 66 | 81 | 72 | |
| 4 | Mon. | 6:20 | 5:40 | 61 | 62 | 56 | |
| 5 | Tues. | 6:19 | 5:41 | 40 | 41 | 31 | very high wind in p.m. |
| 6 | Wed. | 6:17 | 5:43 | 28 | 50 | 46 | |
| 7 | Thurs. | 6:16 | 5:44 | 27 | 39 | 32 | |
| 8 | Fri. | 6:15 | 5:45 | 34 | 50 | 48 | |
| 9 | Sat. | 6:14 | 5:46 | 54 | 50 | 41 | rained last night & a.m., 1.37 |
| 10 | Sun. | 6:12 | 5:48 | 37 | 48 | 42 | |
| 11 | Mon. | 6:11 | 5:49 | 32 | 46 | 36 | |
| 12 | Tues. | 6:10 | 5:50 | 42 | 61 | 57 | sprinkled snow last night |
| 13 | Wed. | 6:09 | 5:51 | 58 | 70 | 57 | apricots in bloom |
| 14 | Thurs. | 6:08 | 5:52 | 40 | 46 | 35 | began snow in p.m. |
| 15 | Fri. | 6:07 | 5:53 | 34 | 46 | 40 | snowed 2.5 last night, nearly all gone |
| 16 | Sat. | 6:06 | 5:54 | 34 | 62 | 60 | |
| 17 | Sun. | 6:04 | 5:56 | 52 | 48 | 42 | |
| 18 | Mon. | 6:03 | 5:57 | 22 | 28 | 26 | began snowing at 2 p.m., 3.75 |
| 19 | Tues. | 6:02 | 5:58 | 25 | 36 | 32 | |
| 20 | Wed. | 6:00 | 6:00 | 29 | 46 | 37 | Vernal Equinox |
| 21 | Thurs. | 5:59 | 6:01 | 34 | 40 | 34 | .5 snow last night |
| 22 | Fri. | 5:58 | 6:02 | 32 | 50 | 40 | sun rose very golden |
| 23 | Sat. | 5:57 | 6:03 | 31 | 60 | 59 | |
| 24 | Sun. | 5:55 | 6:05 | 49 | 58 | 48 | windy |
| 25 | Mon. | 5:54 | 6:06 | 36 | 60 | 52 | shower last night |
| 26 | Tues. | 5:53 | 6:07 | 58 | 70 | 69 | .55 last night, shower this a.m. |
| 27 | Wed. | 5:52 | 6:08 | 66 | 64 | 52 | |
| 28 | Thurs. | 5:51 | 6:09 | 48 | 59 | 48 | |
| 29 | Fri. | 5:50 | 6:10 | 48 | 67 | 58 | |
| 30 | Sat. | 5:49 | 6:11 | 65 | 60 | 50 | |
| 31 | Sun. | 5:48 | 6:12 | 48 | 58 | 50 | Easter Sunday |

## April 1861

Typical spring weather surrounded the historic days during which Richmonders learned of the firing on Fort Sumter, then seceded from the Union. Temperatures taken during the month at 7 a.m., 2 p.m., and 9 p.m. averaged 50.3,
64.1, and 55.7 degrees. Given the importance of the period in Virginia's history, it is worth enumerating the Richmond readings during the eventful mid-
month days, at those same hours of reading: 48, 64, and 60 on the 12th; 58,
72, and 59 on the 13th; 43, 53, and 43 on the 17th; 42, 66, and 64 on the 18th;
52, 53, and 46 on the 19th; 38, 54, and 46 on the 20th; and 45, 72, and 59 on
the 21st. Modest rains blew through Richmond on April 7, 16, and 24. A major
storm dropped 2.25 inches between 4:15 p.m. on April 12 and 8:30 the next
morning. An even larger storm had poured 3.75 inches of rain on Richmond
starting at 7 p.m. on April 10.[20]

On the upper slopes of the Blue Ridge on April 10, a girl wrote in her diary:
"Three days of almost continual rain until today. The sun peeped from the
dark, somber, rolling clouds."[21]

A combination of huge storms wreaked havoc on Fredericksburg and central Virginia. "The weather was horrible" in Caroline County on April 8. The
"heavy storm, which commenced Sunday morning [the 7th], turned to a
regular North Easter" on the night of the 8th and "deluged us with a driving rain." Even heavier downpours followed. "The heavy and continuous rain
of the past few days resulted in a tremendous freshet in the Rappahannock
River, the like of which has not been known since 1814," the *Fredericksburg
News* declared. On April 10 the raging river swept away "panel after panel
of the Falmouth Bridge." That wreckage, "borne down by the impetuous
current, struck the Chatham Bridge 'amidships,' and carried off about one-
third of that structure." Damage to the city gasworks and private mills,
both on the river and along tributary streams, crippled the town. Some families escaped from attic windows in boats. The Rappahannock crested thirty
feet above the high-water mark, twelve inches above the great 1814 flood
and two feet above that of 1847. The Fredericksburg newspaper observed in
wry tribute to the political crisis of the hour that, with the river unbridged
and impassable, "Stafford [County] had *seceded,* by the force of circumstances."[22]

On Saturday, April 13, as news of the firing on Fort Sumter reached Washington, it had "rained all night" in the Northern capital. April 23 in Washington proved to be a "beautiful spring morning," and the 24th continued the "charming weather."[23]

Petersburg's weather on April 13 featured "very mild" temperatures, "showery and sunshiny alternatively," above streets impassably muddy; three days later the Appomattox River there "still continues very high . . . as it rolls turbulently by." "Heavy showers of rain in the morning" dampened a secession celebration in Keysville, Charlotte County, on the 13th. A correspondent in Fincastle reported "pouring down rain" on the evening of April 12.[24]

Well down in Virginia's southwestern tail, citizens in Wytheville rejoiced on April 20 at the prospect of independence from the North and Northerners, but "the late heavy rains" delayed their mail and news, leaving them hungry for information. By April 24, weather on the Atlantic seaboard at Portsmouth had turned "beautiful," as joyous crowds cheered mustering troops.[25]

The rumored crisis in Richmond on the night of April 21–22 involving the gunboat USS *Pawnee* erupted, then collapsed in anti-climax, under a "pale moon" and "twinkling stars."[26]

During just this one month the Georgetown observer divided his ledger's second page, headed "Observations" and intended for comments, with three vertical lines. He obviously intended to provide comments that affected only one-third of any given day. The system did not work particularly well and never appeared again.

## April 1861

| Date | Day of Week | Rich-mond Sunrise | Rich-mond Sunset | D.C. Temperature | | | Other |
|---|---|---|---|---|---|---|---|
| | | | | 7 a.m. | 2 p.m. | 9 p.m. | |
| 1 | Mon. | 5:46 | 6:14 | 44 | 50 | 49 | |
| 2 | Tues. | 5:45 | 6:15 | 50 | 59 | 46 | |
| 3 | Wed. | 5:44 | 6:16 | 42 | 63 | 50 | |
| 4 | Thurs. | 5:43 | 6:17 | 42 | 58 | 48 | rainlike |
| 5 | Fri. | 5:41 | 6:19 | 38 | 65 | 43 | |
| 6 | Sat. | 5:40 | 6:20 | 54 | 62 | 48 | rainlike |
| 7 | Sun. | 5:39 | 6:21 | 48 | 48 | 46 | drizzling all day .30 |
| 8 | Mon. | 5:38 | 6:22 | 45 | 52 | 44 | drizzle then rain |
| 9 | Tues. | 5:36 | 6:24 | 47 | 52 | 47 | 1.12 at 5 p.m. |
| 10 | Wed. | 5:35 | 6:25 | 39 | 44 | 40 | .83, rained all night |
| 11 | Thurs. | 5:34 | 6:26 | 48 | 66 | 52 | |
| 12 | Fri. | 5:33 | 6:27 | 51 | 62 | 57 | began to rain at 2 p.m. |
| 13 | Sat. | 5:32 | 6:28 | 65 | 72 | 56 | 1.51 yesterday & today |
| 14 | Sun. | 5:31 | 6:29 | 56 | 66 | 60 | |
| 15 | Mon. | 5:30 | 6:30 | 52 | 56 | 45 | sprinkled occasionally |
| 16 | Tues. | 5:29 | 6:31 | 46 | 57 | 46 | .85 rain |
| 17 | Wed. | 5:28 | 6:32 | 42 | 42 | 42 | rained thru night |
| 18 | Thurs. | 5:26 | 6:34 | 43 | 64 | 55 | cherries in full bloom |
| 19 | Fri. | 5:25 | 6:35 | 50 | 54 | 46 | |
| 20 | Sat. | 5:24 | 6:36 | 68 | 58 | 49 | |
| 21 | Sun. | 5:23 | 6:37 | 55 | 70 | 56 | |
| 22 | Mon. | 5:22 | 6:38 | 64 | 82 | 66 | |
| 23 | Tues. | 5:21 | 6:39 | 69 | 86 | 66 | |
| 24 | Wed. | 5:20 | 6:40 | 68 | 86 | 62 | very high wind |
| 25 | Thurs. | 5:19 | 6:41 | 62 | 72 | 62 | |
| 26 | Fri. | 5:18 | 6:42 | 56 | 76 | 65 | |
| 27 | Sat. | 5:17 | 6:43 | 62 | 83 | 73 | |
| 28 | Sun. | 5:16 | 6:44 | 73 | 60 | 56 | .17, showers at 10 & 1 |
| 29 | Mon. | 5:15 | 6:45 | 53 | 79 | 61 | |
| 30 | Tues. | 5:14 | 6:46 | 62 | 56 | 56 | high wind at noon |

May 1861

A correspondent in Portsmouth reported on May 2: "The weather here is quite cold . . . though the day is bright and the sun is shining gloriously."[27]

As a volunteer artillery company left Bedford on the morning of May 15, the men reveled in "a beautiful day, no clouds . . . it was just cool enough to be pleasant."[28]

On May 22 in the mountains near Beverly, a Confederate in a newly organizing company, later part of the 31st Virginia Infantry, wrote in his diary, "Morning foggy and cool." The 23rd dawned, he wrote, "cool, frosty." On May 30, a sunny day in Grafton, "the dust nearly suffocates us," the diarist grumbled.[29]

A company from the nascent 13th Virginia Infantry encountered typically fickle May weather when it headed from Harpers Ferry toward Shepherdstown late in the month. The march began on "a beautiful evening" but suddenly ran into what one of the men called "the most terrific hail storm I ever experienced in my life." The terrifying violence of the storm abated almost as suddenly as it had exploded, and the sun promptly reappeared. "The tremendous fall of hail soon melted into water which flowed in every direction, and . . . we marched for several miles in mud and water in many places up to our waists."[30]

At Harrisonburg, after a spectacular sunset on May 29, the weather turned "exceedingly cool" on the 30th.[31]

For some reason, Richmond's weather recorder skipped the first dozen days of the month. The only rain he recorded fell from 8:15 a.m. on May 19 to 9 p.m. on May 20; during that long interval, 1.125 inches accumulated.[32]

A diarist in Richmond described May 12 as "bright and pleasant."[33]

Georgetown's station calculated a month-long total of 2.96 inches of rain, which precisely matches the rainfall it reported for individual days.

## May 1861

| Date | Day of Week | Rich-mond Sunrise | Rich-mond Sunset | D.C. Temperature | | | Other |
|---|---|---|---|---|---|---|---|
| | | | | 7 a.m. | 2 p.m. | 9 p.m. | |
| 1 | Wed. | 5:13 | 6:47 | 56 | 62 | 48 | |
| 2 | Thurs. | 5:12 | 6:48 | 42 | 62 | 46 | |
| 3 | Fri. | 5:11 | 6:49 | 44 | 52 | 46 | drizzling all night |
| 4 | Sat. | 5:10 | 6:50 | 46 | 48 | 48 | .83 rain all last night |
| 5 | Sun. | 5:09 | 6:51 | 48 | 64 | 52 | began to drizzle 3:30 |
| 6 | Mon. | 5:08 | 6:52 | 50 | 58 | 59 | .65 drizzled all day |
| 7 | Tues. | 5:07 | 6:53 | 60 | 72 | 66 | |
| 8 | Wed. | 5:06 | 6:54 | 58 | 66 | 56 | |
| 9 | Thurs. | 5:05 | 6:55 | 58 | 72 | 62 | |
| 10 | Fri. | 5:04 | 6:56 | 58 | 62 | 58 | .32, drizzle began at 11 |
| 11 | Sat. | 5:03 | 6:57 | 58 | 76 | 69 | |
| 12 | Sun. | 5:02 | 6:58 | 58 | 79 | 67 | |
| 13 | Mon. | 5:01 | 6:59 | 66 | 82 | 72 | lightning at 8:30 |
| 14 | Tues. | 5:00 | 7:00 | 70 | 86 | 68 | .14 |
| 15 | Wed. | 4:59 | 7:01 | 68 | 74 | 64 | |
| 16 | Thurs. | 4:58 | 7:02 | 63 | 74 | 59 | |
| 17 | Fri. | 4:57 | 7:03 | 56 | 75 | 60 | much flying dust |
| 18 | Sat. | 4:57 | 7:03 | 57 | 68 | 66 | |
| 19 | Sun. | 4:56 | 7:04 | 56 | 60 | 58 | |
| 20 | Mon. | 4:55 | 7:05 | 57 | 65 | 53 | .65 last night, .17 today |
| 21 | Tues. | 4:55 | 7:05 | 52 | 62 | 58 | |
| 22 | Wed. | 4:54 | 7:06 | 61 | 72 | 57 | |
| 23 | Thurs. | 4:53 | 7:07 | 58 | 73 | 62 | |
| 24 | Fri. | 4:53 | 7:07 | 60 | 72 | 66 | |
| 25 | Sat. | 4:52 | 7:08 | 63 | 82 | 73 | .20 |
| 26 | Sun. | 4:51 | 7:09 | 76 | 86 | 68 | rain & thunder at dusk |
| 27 | Mon. | 4:51 | 7:09 | 76 | 78 | 65 | shower at 10, gusty wind at 3 |
| 28 | Tues. | 4:50 | 7:10 | 60 | 74 | 60 | |
| 29 | Wed. | 4:49 | 7:11 | 62 | 72 | 58 | |
| 30 | Thurs. | 4:49 | 7:11 | 58 | 72 | 62 | |
| 31 | Fri. | 4:48 | 7:12 | 64 | 78 | 68 | |

## June 1861

A soldier marching through western Virginia mountains from Grafton toward Beverly reported in his diary on June 5 a "thunder storm at noon, and rains all evening." The next morning, though, turned "very hot, and men suffer from heat." On the night of the 21st, near Philippi, a "hard thunderstorm" blew down tents and drenched everything. The next day proved "very hot."[34]

A correspondent writing from near Harpers Ferry on June 5 complained: "It has been raining for several days." The next day he reported that the weather had been "so bad for several days that there is but little stirring on the streets." The front obviously stretched all across the Old Dominion: at York-town on the 5th, "the last two or three days it has been raining and wet here."[35]

The same storm that drenched the mountains and Harpers Ferry "poured in torrents all night" of June 5 on a Virginia battery marching from Richmond to West Point. The Richmond artillerists played a key role five days later in the Battle of Big Bethel, under skies bright enough to reflect a "dazzling glitter" from the muskets of approaching Yankees.[36]

Near Manassas, Virginians unaccustomed to field service grumbled about the weather. As they contemplated a march, June 21 proved to be "a scorcher," even at 5 a.m., and the day developed under "a burning sun." The next day also seemed "extremely warm here—much more so, apparently, than at Fairfax Court-House."[37]

"A heavy rain" fell on Confederates camped near Winchester on June 30 and continued the next day "with little cessation." The troops finally received their first issue of tents during the storm, and found in them some shelter from the elements.[38]

A cloth or canvas flap called a havelock, affixed to forage caps and hanging down over the neck and shoulders, offered soldiers protection against sun and rain. This apparently practical device proved more cumbersome than it was worth, however, and quickly disappeared from the field. The name derived from supposed origins in the armies of Sir Henry Havelock, a British officer recently of note in quelling the Sepoy Mutiny in India, a locale with a climate even more intemperate than Virginia's summers. (Benson J. Lossing, *Pictorial History of the Civil War in the United States of America*, 3 vols. [Philadelphia: G. W. Childs, 1866–68], 1:575.)

June 1861

| Date | Day of Week | Rich-mond Sunrise | Rich-mond Sunset | D.C. Temperature | | | Other |
|------|------|------|------|------|------|------|------|
| | | | | 7 a.m. | 2 p.m. | 9 p.m. | |
| 1 | Sat. | 4:48 | 7:12 | 76 | 79 | 69 | |
| 2 | Sun. | 4:47 | 7:13 | 73 | 72 | 78 | evening .15 |
| 3 | Mon. | 4:47 | 7:13 | 73 | 88 | 78 | |
| 4 | Tues. | 4:46 | 7:14 | 72 | 66 | 66 | misty at 7, occ. rain |
| 5 | Wed. | 4:46 | 7:14 | 65 | 65 | 67 | .82, drizzly at 7 |
| 6 | Thurs. | 4:45 | 7:15 | 56 | 69 | 64 | rained all night |
| 7 | Fri. | 4:45 | 7:15 | 65 | 73 | 68 | first firefly seen this evening |
| 8 | Sat. | 4:44 | 7:16 | 67 | 70 | 67 | .2 at noon, +.98 |
| 9 | Sun. | 4:44 | 7:16 | 67 | 78 | 68 | |
| 10 | Mon. | 4:44 | 7:16 | 70 | 86 | 76 | |
| 11 | Tues. | 4:44 | 7:16 | 75 | 88 | 77 | |
| 12 | Wed. | 4:43 | 7:17 | 76 | 90 | 81 | |
| 13 | Thurs. | 4:43 | 7:17 | 80 | 82 | 78 | |
| 14 | Fri. | 4:43 | 7:17 | 66 | 82 | 77 | |
| 15 | Sat. | 4:43 | 7:17 | 78 | 96 | 86 | |
| 16 | Sun. | 4:42 | 7:18 | 84 | 86 | 75 | |
| 17 | Mon. | 4:42 | 7:18 | 66 | 72 | 68 | slight shower last night |
| 18 | Tues. | 4:42 | 7:18 | 66 | 78 | 70 | |
| 19 | Wed. | 4:42 | 7:18 | 69 | 84 | 72 | |
| 20 | Thurs. | 4:42 | 7:18 | 72 | 82 | 76 | slight shower at day-break |
| 21 | Fri. | 4:42 | 7:18 | 75 | 94 | 87 | Summer Solstice |
| 22 | Sat. | 4:42 | 7:18 | 76 | 90 | 70 | slight shower last night with wind |
| 23 | Sun. | 4:42 | 7:18 | 80 | 91 | 78 | |
| 24 | Mon. | 4:42 | 7:18 | 76 | 96 | 74 | |
| 25 | Tues. | 4:42 | 7:18 | 69 | 88 | 78 | |
| 26 | Wed. | 4:42 | 7:18 | 76 | 90 | 77 | shower last night & this evening |
| 27 | Thurs. | 4:43 | 7:17 | 76 | 90 | 73 | |
| 28 | Fri. | 4:43 | 7:17 | 71 | 87 | 78 | |
| 29 | Sat. | 4:43 | 7:17 | 71 | 81 | 71 | slight drizzle thru night |
| 30 | Sun. | 4:43 | 7:17 | 70 | 66 | 67 | gentle rain, clear at 6 p.m., fireflies un-usually numerous |

July 1861

A Georgian marching in the Shenandoah Valley on July 17 drolly described the effects of the weather: "The sun shone warmly, and the cruel government had provided us with no umbrellas."[39]

The day of the war's first major battle, July 21, unfolded with weather "bright and beautiful," under a strikingly blue sky.[40]

The 1st Texas Infantry did not participate in the Battle of First Manassas, but hurried northward from Richmond on the Richmond, Fredericksburg & Potomac Railroad in case they might be needed as reinforcements. The regiment ran into "pouring rain" that evening of July 21 and suffered a deadly wreck when derailed by a washed-out culvert.[41]

On the 22nd a Confederate officer near Manassas wrote, "it rained all day, much of the time very hard."[42] A quartermaster described the precipitation as falling "in torrents," and a Marylander declared emphatically that "the day was the most damp and dismal one as far as the weather was concerned that I experienced while in the army."[43]

Far to the west of the plains of Manassas, Southerners skirmishing near Cheat Mountain also had "rain all day," then on July 23, faced a "night so cold almost freezing."[44]

Atop the Blue Ridge west of Manassas, July 25 "dawned unclouded, bright and beautiful."[45]

In Richmond, temperatures recorded at 7 a.m., 2 p.m., and 9 p.m. averaged 68.8, 84.7, and 73.2 degrees for the month. On July 18 and 21, while the armies fought at Manassas, Richmond readings at those three hours were 72, 92, and 76 on the 18th and 70, 84, and 75 on the 21st. Rain fell in Richmond during eight July days, totaling 8.25 inches, five of them on the notoriously wet 22nd.[46]

## July 1861

| Date | Day of Week | Rich-mond Sunrise | Rich-mond Sunset | D.C. Temperature | | | Other |
|------|------|------|------|------|------|------|------|
| | | | | 7 a.m. | 2 p.m. | 9 p.m. | |
| 1 | Mon. | 4:43 | 7:17 | 66 | 81 | 64 | 1.34, high wind |
| 2 | Tues. | 4:44 | 7:16 | 64 | 75 | 66 | |
| 3 | Wed. | 4:44 | 7:16 | 67 | 83 | 70 | |
| 4 | Thurs. | 4:44 | 7:16 | 68 | 82 | 70 | |
| 5 | Fri. | 4:45 | 7:15 | 69 | 83 | 76 | |
| 6 | Sat. | 4:45 | 7:15 | 74 | 78 | 76 | .65 last night |
| 7 | Sun. | 4:45 | 7:15 | 80 | 90 | 79 | slight shower overnight |
| 8 | Mon. | 4:46 | 7:14 | 78 | 95 | 78 | heavy clouds at 4, but clear by dark |
| 9 | Tues. | 4:46 | 7:14 | 82 | 96 | 74 | 1.05, heavy shower 6–7 p.m. |
| 10 | Wed. | 4:47 | 7:13 | 77 | 90 | 67 | .94, rain 6–9 p.m. |
| 11 | Thurs. | 4:47 | 7:13 | 73 | 81 | 78 | .16, slight shower at 6 |
| 12 | Fri. | 4:48 | 7:12 | 70 | 82 | 69 | |
| 13 | Sat. | 4:48 | 7:12 | 64 | 75 | 65 | last night .14 |
| 14 | Sun. | 4:49 | 7:11 | 66 | 74 | 66 | |
| 15 | Mon. | 4:50 | 7:10 | 67 | 82 | 71 | |
| 16 | Tues. | 4:50 | 7:10 | 74 | 82 | 72 | |
| 17 | Wed. | 4:51 | 7:09 | 72 | 86 | 70 | |
| 18 | Thurs. | 4:52 | 7:08 | 74 | 90 | 80 | |
| 19 | Fri. | 4:52 | 7:08 | 74 | 82 | 80 | rainlike at 7 a.m. |
| 20 | Sat. | 4:53 | 7:07 | 75 | 90 | 77 | sprinkled last night |
| 21 | Sun. | 4:54 | 7:06 | 68 | 80 | 73 | |
| 22 | Mon. | 4:55 | 7:05 | 70 | 66 | 63 | drizzly at 7, rain all day |
| 23 | Tues. | 4:55 | 7:05 | 68 | 80 | 65 | 1.45 yest. & last night |
| 24 | Wed. | 4:56 | 7:04 | 77 | 80 | 68 | |
| 25 | Thurs. | 4:57 | 7:03 | 70 | 82 | 66 | |
| 26 | Fri. | 4:57 | 7:03 | 66 | 86 | 76 | |
| 27 | Sat. | 4:58 | 7:02 | 72 | 81 | 71 | |
| 28 | Sun. | 4:59 | 7:01 | 69 | 88 | 73 | .49, showery 6–7 p.m. |
| 29 | Mon. | 4:59 | 7:01 | 74 | 86 | 75 | |
| 30 | Tues. | 5:00 | 7:00 | 66 | 90 | 74 | |
| 31 | Wed. | 5:01 | 6:59 | 77 | 93 | 82 | |

# August 1861

This month featured "unfavorable weather," in the words of a Georgian camped three miles above Manassas at Camp Bartow. Colonel George T. "Tige" Anderson, commander of the 11th Georgia, reported a "severe storm, rain, and wind" on August 9 and at mid-month grumbled of "rain all day and for three days dull misty damp, cloudy and disagreeable."[47]

A North Carolinian described the steady spell of bad weather mid-month in a letter dated August 23: "It has been raining here for nearly a week, and it is tolerably cool. This morning was very cool and chilly."[48]

Despite the summer date, a Virginian camped on the Greenbrier River in the western mountains wrote in his diary on August 30, "night bitter cold, no sleep."[49]

The rainfall in the mountains continued so relentlessly that it destroyed all semblance of careful military operations. A Virginia soldier who reached the front there on August 6 wrote: "It was indeed a fearful summer. We were camped on Valley Mountain 43 days, and it rained 37 days out of that number." A waggish member of the 21st Virginia Infantry insisted that "it rained thirty-*two* days in August." General R. E. Lee could make no headway against either the Yankees or the weather under these frustrating circumstances.[50]

A refugee quartered near Fredericksburg reported these weather notes. 1st: "Clear and cool." 14th: "Lovely day." 19th: "It has been raining for days." 23rd: "Another terrific rain storm." 28th: "How refreshing to see the sunshine once more."[51] Rain returned "hard all day" on August 29, long enough and hard enough to make travel difficult near Fredericksburg, Betty Herndon Maury wrote in her diary. Betty's brilliant father, Matthew Fontaine Maury, had earned before the war a reputation for several scientific endeavors, including in the meteorological field, so his daughter's attention to the weather came naturally. Her comments will appear below through June 1862.[52]

The Georgetown station calculated August's total rainfall as an amazingly high 6.91 inches, which precisely matches the individual totals reported on individual days.

The dreadful weather in western Virginia's mountains during August 1861 thwarted the efforts of General Robert E. Lee to bring order from chaos. Perhaps the "thirty-*two* days" of rain that month contributed to turning his dark beard, seen in this drawing, into the more familiar gray hair of future months. (Robert Alonzo Brock, ed., *Gen. Robert Edward Lee: Soldier, Citizen, and Christian Patriot* [Richmond: Royal Publishing Co., 1897], 174.)

Betty Herndon Maury, daughter of Matthew Fontaine Maury, inherited her father's interest in the weather and recorded details in her published diary. (Betty H. Maury, *The Confederate Diary of Betty Herndon Maury* [Washington, D.C.: privately printed, 1938], frontispiece.)

## August 1861

| Date | Day of Week | Rich- mond Sunrise | Rich- mond Sunset | D.C. Temperature | | | Other |
|---|---|---|---|---|---|---|---|
| | | | | 7 a.m. | 2 p.m. | 9 p.m. | |
| 1 | Thurs. | 5:02 | 6:58 | 74 | 86 | 80 | last night 1.0 |
| 2 | Fri. | 5:03 | 6:57 | 81 | 86 | 80 | .82, heavy fog |
| 3 | Sat. | 5:04 | 6:56 | 77 | 86 | 82 | |
| 4 | Sun. | 5:05 | 6:55 | 82 | 95 | 82 | |
| 5 | Mon. | 5:06 | 6:54 | 81 | 84 | 79 | |
| 6 | Tues. | 5:07 | 6:53 | 72 | 90 | 77 | .14 last night |
| 7 | Wed. | 5:08 | 6:52 | 76 | 92 | 80 | |
| 8 | Thurs. | 5:09 | 6:51 | 80 | 93 | 81 | very warm, sultry |
| 9 | Fri. | 5:10 | 6:50 | 77 | 83 | 77 | misty 7 a.m., showery 7 p.m. |
| 10 | Sat. | 5:11 | 6:49 | 74 | 90 | 83 | .24 last night & a.m. |
| 11 | Sun. | 5:12 | 6:48 | 82 | 89 | 76 | heavy rain began 5 p.m. |
| 12 | Mon. | 5:13 | 6:47 | 76 | 76 | 75 | 2.55 overnight |
| 13 | Tues. | 5:14 | 6:46 | 76 | 73 | 64 | .67 last night |
| 14 | Wed. | 5:15 | 6:45 | 63 | 69 | 66 | |
| 15 | Thurs. | 5:16 | 6:44 | 62 | 75 | 66 | |
| 16 | Fri. | 5:17 | 6:43 | 65 | 70 | 64 | raining from 3 p.m. |
| 17 | Sat. | 5:18 | 6:42 | 66 | 74 | 72 | drizzled all day |
| 18 | Sun. | 5:19 | 6:41 | 72 | 78 | 75 | drizzled all day |
| 19 | Mon. | 5:20 | 6:40 | 74 | 76 | 74 | rained occasionally |
| 20 | Tues. | 5:21 | 6:39 | 70 | 80 | 68 | .89 yesterday & last night |
| 21 | Wed. | 5:22 | 6:38 | 65 | 77 | 76 | |
| 22 | Thurs. | 5:23 | 6:37 | 71 | 84 | 74 | showery all day |
| 23 | Fri. | 5:24 | 6:36 | 66 | 87 | 71 | |
| 24 | Sat. | 5:25 | 6:35 | 63 | 80 | 67 | |
| 25 | Sun. | 5:27 | 6:33 | 66 | 82 | 69 | first katydid heard this season |
| 26 | Mon. | 5:28 | 6:32 | 68 | 80 | 74 | |
| 27 | Tues. | 5:29 | 6:31 | 68 | 76 | 72 | sprinkled thru p.m. |
| 28 | Wed. | 5:30 | 6:30 | 71 | 77 | 72 | drizzling all day |
| 29 | Thurs. | 5:31 | 6:29 | 72 | 76 | 70 | raining all day |
| 30 | Fri. | 5:32 | 6:28 | 69 | 82 | 73 | .60 yesterday & today |
| 31 | Sat. | 5:33 | 6:27 | 65 | 77 | 64 | |

September 1861

A soldier serving with the 22nd Virginia Infantry in the mountains of western Virginia remembered "the equinoctial storms of September, 1861" with undisguised disgust: "The deluge of rain flooded the mountains, and the overflowing rivulets, numerous and terrific, cut off our supplies."[53] A diary entry recorded on the 27th, near Cheat Mountain: "Terrible storm, wind and rain. Tents capsized; river past fording. Nothing for our horses to eat."[54]

The mountains absorbed the rain, only a little pushing on into central Virginia. Richmond measured only 0.88 inches of rain in four small storms on September 4–5, 11–12, 21, and 26–27. The highest reading all month in Virginia's capital city was 89 at 2 p.m. on the 17th, the lowest 46.5 at 7 a.m. on the 29th. Average temperatures for the month at 7 a.m., 2 p.m., and 9 p.m. were 63.5, 79.4, and 69.8 degrees.[55]

"Bright, warm and delightful weather" shone on Fairfax County on September 29.[56]

September 1861

| Date | Day of Week | Rich-mond Sunrise | Rich-mond Sunset | D.C. Temperature | | | Other |
|------|------|------|------|------|------|------|------|
| | | | | 7 a.m. | 2 p.m. | 9 p.m. | |
| 1 | Sun. | 5:34 | 6:26 | 61 | 77 | 66 | |
| 2 | Mon. | 5:35 | 6:25 | 67 | 80 | 73 | |
| 3 | Tues. | 5:36 | 6:24 | 70 | 89 | 80 | |
| 4 | Wed. | 5:38 | 6:22 | 72 | 84 | 74 | slight shower last night |
| 5 | Thurs. | 5:39 | 6:21 | 65 | 86 | 67 | shower last night, rain all day |
| 6 | Fri. | 5:40 | 6:20 | 68 | 82 | 74 | .42 last night & yest. |
| 7 | Sat. | 5:41 | 6:19 | 67 | 84 | 72 | |
| 8 | Sun. | 5:42 | 6:18 | 68 | 76 | 68 | |
| 9 | Mon. | 5:44 | 6:16 | 68 | 80 | 71 | |
| 10 | Tues. | 5:45 | 6:15 | 68 | 84 | 72 | saw fireflies this evening |
| 11 | Wed. | 5:46 | 6:14 | 72 | 82 | 78 | |
| 12 | Thurs. | 5:47 | 6:13 | 70 | 80 | 70 | .91 last night |
| 13 | Fri. | 5:48 | 6:12 | 67 | 86 | 74 | |
| 14 | Sat. | 5:49 | 6:11 | 67 | 84 | 74 | |
| 15 | Sun. | 5:50 | 6:10 | 70 | 89 | 76 | |
| 16 | Mon. | 5:51 | 6:09 | 74 | 82 | 66 | |
| 17 | Tues. | 5:53 | 6:07 | 73 | 86 | 69 | .70, heavy shower 4:30 p.m. |
| 18 | Wed. | 5:54 | 6:06 | 66 | 76 | 69 | |
| 19 | Thurs. | 5:55 | 6:05 | 70 | 82 | 70 | |
| 20 | Fri. | 5:56 | 6:04 | 67 | 83 | 78 | |
| 21 | Sat. | 5:58 | 6:02 | 75 | 74 | 68 | .34, rain began at 5 p.m. |
| 22 | Sun. | 5:59 | 6:01 | 64 | 66 | 58 | fireflies this evening |
| 23 | Mon. | 6:00 | 6:00 | 55 | 76 | 68 | Autumnal Equinox |
| 24 | Tues. | 6:01 | 5:59 | 60 | 78 | 66 | heavy dew |
| 25 | Wed. | 6:03 | 5:57 | 56 | 78 | 65 | heavy dew |
| 26 | Thurs. | 6:04 | 5:56 | 65 | 78 | 70 | slight rain thru night |
| 27 | Fri. | 6:05 | 5:55 | 66 | 75 | 66 | rain began 2 p.m., heavy wind |
| 28 | Sat. | 6:07 | 5:53 | 68 | 70 | 56 | |
| 29 | Sun. | 6:08 | 5:52 | 50 | 71 | 52 | |
| 30 | Mon. | 6:09 | 5:51 | 56 | 72 | 54 | |

## October 1861

The seasons changed during the desultory campaigning around Cheat Mountain in the western hills. An October 7 diary entry affords a mixed review: "Gloomy day—leaves are falling, the forests beautiful."[57]

On October 10, Virginia artillerists went to Cape Henry to fire on a Yankee gunboat "blown aground in a storm the night before."[58]

October in Richmond passed with cool weather and only moderate rainfall. Average readings for the month at 7 a.m., 2 p.m., and 9 p.m. were 56.5, 69.9, and 61.4. On October 21, while men fought northward on the banks of the Potomac River at Ball's Bluff, the temperature in Richmond at those three hours was 55, 62, and 57, with rain that night. Rain fell during seven October days in Richmond, but obviously only lightly, as the month's total only reached 1.25 inches.[59]

The much heavier rainfall in Washington for October 1861 totaled 3.93 inches, both in the observer's final report and in the accumulated totals for individual days.

For the first time during the war, thoughtful officers began to recognize the need to plan for cold-weather gear for the troops. From a camp near Manassas, a Georgian wrote on October 17: "The men have received no clothing yet. . . . We have already had days cold enough for greatcoats." Northerners across the lines complained of the same cold, but on the 25th a Rhode Island regiment received a box of blankets sent from home, ending their discomfort.[60]

The major of the 24th Virginia spent the night of October 22, which was to have been his wedding night had he managed to get a furlough, near Union Mills "on picket duty in the pine woods in a driving rain without fire or tent."[61]

From the camp of a North Carolina regiment near Manassas, a letter of October 24 reported: "We have had frost for several nights and it is already beginning to turn very cold."[62]

## October 1861

| Date | Day of Week | Rich-mond Sunrise | Rich-mond Sunset | D.C. Temperature | | | Other |
|------|------|------|------|------|------|------|------|
| | | | | 7 a.m. | 2 p.m. | 9 p.m. | |
| 1 | Tues. | 6:10 | 5:50 | 52 | 70 | 62 | |
| 2 | Wed. | 6:11 | 5:49 | 64 | 78 | 65 | began raining gently at 10 |
| 3 | Thurs. | 6:12 | 5:48 | 61 | 82 | 70 | foggy, smoky |
| 4 | Fri. | 6:13 | 5:47 | 62 | 86 | 73 | |
| 5 | Sat. | 6:14 | 5:46 | 71 | 88 | 75 | |
| 6 | Sun. | 6:16 | 5:44 | 74 | 82 | 80 | |
| 7 | Mon. | 6:17 | 5:43 | 76 | 85 | 63 | began rain at 5 p.m. |
| 8 | Tues. | 6:18 | 5:42 | 63 | 68 | 60 | 2.2 last night |
| 9 | Wed. | 6:19 | 5:41 | 57 | 66 | 62 | |
| 10 | Thurs. | 6:21 | 5:39 | 61 | 62 | 58 | rained slightly all day |
| 11 | Fri. | 6:22 | 5:38 | 60 | 73 | 67 | .14 yesterday and last night |
| 12 | Sat. | 6:23 | 5:37 | 56 | 66 | 58 | .40, rained thru night |
| 13 | Sun. | 6:24 | 5:36 | 57 | 73 | 54 | |
| 14 | Mon. | 6:26 | 5:34 | 45 | 68 | 56 | |
| 15 | Tues. | 6:27 | 5:33 | 48 | 74 | 60 | |
| 16 | Wed. | 6:28 | 5:32 | 51 | 68 | 63 | |
| 17 | Thurs. | 6:29 | 5:31 | 64 | 74 | 69 | |
| 18 | Fri. | 6:30 | 5:30 | 70 | 78 | 70 | .65 rain last night, began 6 p.m. |
| 19 | Sat. | 6:31 | 5:29 | 68 | 78 | 71 | heavy fog and dew |
| 20 | Sun. | 6:32 | 5:28 | 64 | 66 | 58 | |
| 21 | Mon. | 6:33 | 5:27 | 51 | 60 | 54 | |
| 22 | Tues. | 6:34 | 5:26 | 54 | 63 | 60 | rained all day |
| 23 | Wed. | 6:36 | 5:24 | 58 | 62 | 46 | .54 yesterday & last night |
| 24 | Thurs. | 6:37 | 5:23 | 42 | 57 | 45 | |
| 25 | Fri. | 6:38 | 5:22 | 39 | 58 | 54 | heavy white frost |
| 26 | Sat. | 6:39 | 5:21 | 52 | 64 | 62 | |
| 27 | Sun. | 6:40 | 5:20 | 59 | 58 | 52 | slight shower last night |
| 28 | Mon. | 6:41 | 5:19 | 36 | 58 | 54 | slight frost |
| 29 | Tues. | 6:43 | 5:17 | 41 | 66 | 48 | |
| 30 | Wed. | 6:44 | 5:16 | 48 | 68 | 58 | rainlike this evening |
| 31 | Thurs. | 6:45 | 5:15 | 44 | 62 | 50 | smoky this a.m. |

# November 1861

A Virginian camped near Manassas awoke at 4 a.m. on November 2 when his tent blew down, leaving him "exposed to a heavy rain." November 3 "was calm and beautiful in marked contrast with the storm that had just preceded it." Colonel Anderson of the 11th Georgia described the November 2 weather as "a very severe storm of rain and wind." On the 16th, Anderson wrote in his diary of another "severe storm of wind and cold" that left "snow visible on peaks of the Blue Ridge."[63]

Rain fell on Fredericksburg on November 5 and 10.[64]

Miserable weather continued to dog Confederate troops trying to operate in the western mountains. Colonel Robert H. Hatton of Tennessee wrote from Greenbrier County to his wife on November 2: "It is raining. Rained all last night. Wind blew hard," knocking over tents. From Warm Springs, Hatton wrote on November 27: "We are in the midst of real Winter, here—snow, frosts, and ice in abundance."[65]

A Georgia officer stationed near Manassas wrote on the morning of November 23: "A little sleet and rain—a bad prospect for the poor soldier."[66]

The Georgetown observer noted sighting a meteor on November 7: "About 7 p.m. a very bright meteor (falling star) descended from about 60 degrees above the Western Horizon to nearly the earth." The November sheet figured rainfall at 3.90 inches in Georgetown, which for some reason (probably sheer mathematical ineptitude) falls .09 inches below the sum of the reports for individual days.[67]

November 1861

| Date | Day of Week | Rich-mond Sunrise | Rich-mond Sunset | D.C. Temperature | | | Other |
|------|------|------|------|------|------|------|------|
| | | | | 7 a.m. | 2 p.m. | 9 p.m. | |
| 1 | Fri. | 6:46 | 5:14 | 41 | 60 | 55 | raining at 9 p.m. |
| 2 | Sat. | 6:47 | 5:13 | 62 | 66 | 63 | 2.0, rained all night |
| 3 | Sun. | 6:48 | 5:12 | 63 | 60 | 52 | |
| 4 | Mon. | 6:49 | 5:11 | 56 | 60 | 49 | |
| 5 | Tues. | 6:50 | 5:10 | 40 | 62 | 49 | white frost, heavy fog |
| 6 | Wed. | 6:51 | 5:09 | 40 | 55 | 52 | rained occasionally |
| 7 | Thurs. | 6:52 | 5:08 | 47 | 54 | 46 | |
| 8 | Fri. | 6:53 | 5:07 | 35 | 61 | 54 | .17 last night |
| 9 | Sat. | 6:54 | 5:06 | 48 | 50 | 46 | heavy white frost |
| 10 | Sun. | 6:55 | 5:05 | 40 | 52 | 48 | |
| 11 | Mon. | 6:56 | 5:04 | 49 | 66 | 55 | |
| 12 | Tues. | 6:57 | 5:03 | 42 | 59 | 50 | |
| 13 | Wed. | 6:58 | 5:02 | 44 | 60 | 58 | |
| 14 | Thurs. | 6:59 | 5:01 | 50 | 60 | 54 | began raining at 2, heavier at 9 |
| 15 | Fri. | 7:00 | 5:00 | 45 | 60 | 44 | .51 last night & today |
| 16 | Sat. | 7:01 | 4:59 | 38 | 60 | 39 | |
| 17 | Sun. | 7:01 | 4:59 | 42 | 49 | 43 | |
| 18 | Mon. | 7:02 | 4:58 | 39 | 53 | 43 | |
| 19 | Tues. | 7:03 | 4:57 | 34 | 54 | 39 | |
| 20 | Wed. | 7:04 | 4:56 | 33 | 48 | 47 | rainlike |
| 21 | Thurs. | 7:05 | 4:55 | 40 | 56 | 48 | .51 last night & this a.m. |
| 22 | Fri. | 7:06 | 4:54 | 36 | 54 | 49 | |
| 23 | Sat. | 7:07 | 4:53 | 45 | 48 | 43 | |
| 24 | Sun. | 7:07 | 4:53 | 37 | 46 | 37 | at 8 p.m. began to snow |
| 25 | Mon. | 7:08 | 4:52 | 33 | 44 | 39 | snow covered the roof |
| 26 | Tues. | 7:09 | 4:51 | 34 | 46 | 37 | |
| 27 | Wed. | 7:09 | 4:51 | 32 | 41 | 46 | at 7 murky horizon |
| 28 | Thurs. | 7:10 | 4:50 | 34 | 53 | 48 | heavy white frost |
| 29 | Fri. | 7:11 | 4:49 | 54 | 68 | 62 | murky, overcast |
| 30 | Sat. | 7:11 | 4:49 | 48 | 58 | 40 | .80 last night, windy |

December 1861

At Ashby's Gap, December opened with "a dull gloomy day."[68]

On December 3 near Fredericksburg, "Snowing steadily, big, soft white flakes. . . . It is very beautiful." Temperatures remained "bitterly cold" for a time, moderated on the 16th, and by the 20th had become "frightfully cold."[69]

A Tennessee colonel who wrote to his wife from Staunton on December 16 rejoiced in a "bright and warm" day. He described Christmas in Strasburg as lovely: "The sun is out in his brightness this morning. Yesterday, and night before last, the weather was terrible; now the air is greatly warmer." His troops marched on to Winchester the next two days through good weather.[70]

Sergeant Major Randolph Barton of the 33rd Virginia Infantry described "a warm, gentle rain" in the Shenandoah Valley on the last day of the year and into the night.[71]

The comments under "Observer's Remarks" in the December 1861 table are particularly reliable, because the Georgetown weather observer recorded dates on his second page as well as the first, removing any question about correct alignment.

## December 1861

| Date | Day of Week | Rich- mond Sunrise | Rich- mond Sunset | D.C. Temperature | | | Other |
|---|---|---|---|---|---|---|---|
| | | | | 7 a.m. | 2 p.m. | 9 p.m. | |
| 1 | Sun. | 7:12 | 4:48 | 40 | 48 | 48 | quite snowlike |
| 2 | Mon. | 7:12 | 4:48 | 37 | 38 | 35 | snowlike all day |
| 3 | Tues. | 7:13 | 4:47 | 24 | 35 | 30 | |
| 4 | Wed. | 7:13 | 4:47 | 22 | 36 | 34 | |
| 5 | Thurs. | 7:14 | 4:46 | 29 | 50 | 40 | |
| 6 | Fri. | 7:14 | 4:46 | 34 | 58 | 42 | |
| 7 | Sat. | 7:15 | 4:45 | 36 | 50 | 40 | |
| 8 | Sun. | 7:15 | 4:45 | 38 | 62 | 48 | |
| 9 | Mon. | 7:15 | 4:45 | 42 | 70 | 56 | |
| 10 | Tues. | 7:16 | 4:44 | 49 | 58 | 60 | slight fog, stones moist |
| 11 | Wed. | 7:16 | 4:44 | 64 | 52 | 38 | very windy & dusty |
| 12 | Thurs. | 7:16 | 4:44 | 28 | 42 | 32 | heavy white frost |
| 13 | Fri. | 7:17 | 4:43 | 22 | 42 | 36 | |
| 14 | Sat. | 7:17 | 4:43 | 27 | 52 | 41 | heavy white frost |
| 15 | Sun. | 7:17 | 4:43 | 44 | 51 | 35 | white frost |
| 16 | Mon. | 7:17 | 4:43 | 28 | 58 | 47 | |
| 17 | Tues. | 7:18 | 4:42 | 46 | 56 | 46 | heavy white frost |
| 18 | Wed. | 7:18 | 4:42 | 34 | 57 | 40 | white frost |
| 19 | Thurs. | 7:18 | 4:42 | 32 | 58 | 46 | white frost |
| 20 | Fri. | 7:18 | 4:42 | 40 | 56 | 44 | white frost |
| 21 | Sat. | 7:18 | 4:42 | 28 | 36 | 28 | Winter Solstice |
| 22 | Sun. | 7:18 | 4:42 | 32 | 40 | 36 | no frost, rain 8 p.m. |
| 23 | Mon. | 7:18 | 4:42 | 34 | 40 | 36 | drizzled a.m., snow shower 2:30 |
| 24 | Tues. | 7:18 | 4:42 | 32 | 36 | — | high wind thru night |
| 25 | Wed. | 7:18 | 4:42 | 30 | 42 | 34 | |
| 26 | Thurs. | 7:18 | 4:42 | 31 | 43 | 34 | |
| 27 | Fri. | 7:17 | 4:43 | 46 | 38 | 32 | |
| 28 | Sat. | 7:17 | 4:43 | 41 | 46 | 46 | |
| 29 | Sun. | 7:17 | 4:43 | 45 | 53 | 44 | snowlike |
| 30 | Mon. | 7:17 | 4:43 | 50 | 57 | 44 | white frost |
| 31 | Tues. | 7:17 | 4:43 | 44 | 59 | 38 | |

# Chapter 3

# 1862

## January 1862

One of the war's military operations in Virginia most affected by weather began on January 1, when General Thomas J. "Stonewall" Jackson pushed west from Winchester toward Romney on "a bright, warm day, with a touch of spring in the air." A diarist a few miles south of Winchester enthused over a New Year's Day that offered "Summer's sunshine" and "Spring's balmy breath."[1]

A weather station at Cumberland, Maryland, in the mountains toward which Jackson was moving, provides solid data on the weather. Unfortunately, the Cumberland ledger only recorded temperature and conditions once daily, at 7 a.m. (presumably near the coldest point of each day). These data show the drastic and abrupt change of which so many soldiers complained; but they also reveal that the bitter cold did not plunge below zero, as most participants insisted, based on their very real suffering. The following 7 a.m. readings from Cumberland cover the heart of the campaign: 55 and windy on the 1st; 27 and fair on the 2nd; 18 and fair on the 3rd; 24 and dull on the 4th; 8 and fair on the 5th; 20 and 4 inches of snow on the 6th; 26 and fair on the 7th; 34 and dull on the 9th; 39 on the 10th; 30 on the 11th; 38 and fair on the 12th; 21 and snow on the 13th; 22 and dull on the 14th; and 23 and dull on the 15th.[2]

A colonel in a contemporary letter about Jackson's frozen campaign toward Romney called the venture a "very heavy and wearisome march." His rank doubtless gave him better opportunities than most to cope with the "exceedingly cold" weather, but he still shuddered at the ordeal: "The ground being covered with a thick coating of snow. . . . I had not slept any for two or three nights." Eventually he adjusted: "For a week past, I have, each night, cleared away a place in the snow, to lie down, and . . . have slept . . . as soundly as I *ever* slept."[3]

The dreadful weather en route to Romney, with roads "as slick as glass" under cursing soldiers' feet, reminded a Georgian of a rustic rhyme from his youth: "First she blew, / Then she snew, / And then she thew, / And then she friz."[4]

Confederates encamped near Manassas avoided extended marches in the same cold weather, but of course its effects extended to their longitude. "We have a hard time of it here now," a North Carolina soldier wrote on January 16. "The ground is covered with snow and then a sleet over that, and it is nearly as cold as the frozen regions, the winds . . . blow around us like a regular hurricane."[5]

Soldiers in the camps clustered around Manassas emulated their counterparts through the ages by taking fun where it could be found and at whatever risks. An officer from Georgia wrote on January 8: "It is raining. There was a snow last week, and it is on the ground yet. . . . We have had fun snowballing and sliding downhill on planks since it snowed. Some of the boys go skating on a neighboring mill pond. Unfortunately, one poor fellow broke through the other day and was drowned."[6]

## January 1862

| Date | Day of Week | Rich-mond Sunrise | Rich-mond Sunset | D.C. Temperature | | | Other |
|------|------|------|------|------|------|------|------|
| | | | | 7 a.m. | 2 p.m. | 9 p.m. | |
| 1 | Wed. | 7:16 | 4:44 | 50 | — | — | lurid sky |
| 2 | Thurs. | 7:16 | 4:44 | — | — | 29 | windy all night |
| 3 | Fri. | 7:15 | 4:45 | 24 | — | 27 | |
| 4 | Sat. | 7:15 | 4:45 | 22 | 25 | 23 | |
| 5 | Sun. | 7:15 | 4:45 | 13 | 27 | 24 | began sleeting at dark |
| 6 | Mon. | 7:14 | 4:46 | 23 | 29 | 28 | began snowing about 10 |
| 7 | Tues. | 7:14 | 4:46 | 18 | 36 | 27 | spitting snow at 8 |
| 8 | Wed. | 7:13 | 4:47 | — | 38 | 36 | |
| 9 | Thurs. | 7:12 | 4:48 | 36 | 46 | 38 | sprinkling at 9 p.m. |
| 10 | Fri. | 7:12 | 4:48 | 36 | 45 | 43 | drizzling all night |
| 11 | Sat. | 7:11 | 4:49 | 37 | 44 | — | |
| 12 | Sun. | 7:11 | 4:49 | 40 | 66 | 63 | sprinkling at 9 p.m. |
| 13 | Mon. | 7:10 | 4:50 | 46 | 36 | 31 | |
| 14 | Tues. | 7:09 | 4:51 | 21 | 38 | 23 | high wind all last night |
| 15 | Wed. | 7:09 | 4:51 | 23 | 30 | 36 | |
| 16 | Thurs. | 7:08 | 4:52 | 35 | 39 | 28 | sleeted all day to 9 p.m. |
| 17 | Fri. | 7:07 | 4:53 | 22 | 42 | 38 | |
| 18 | Sat. | 7:06 | 4:54 | 35 | 42 | 42 | |
| 19 | Sun. | 7:06 | 4:54 | 39 | 51 | 50 | |
| 20 | Mon. | 7:05 | 4:55 | 39 | 44 | 41 | |
| 21 | Tues. | 7:04 | 4:56 | 36 | 34 | 33 | |
| 22 | Wed. | 7:03 | 4:57 | 31 | 35 | 35 | |
| 23 | Thurs. | 7:03 | 4:57 | 32 | 35 | 35 | |
| 24 | Fri. | 7:02 | 4:58 | 33 | 36 | 33 | |
| 25 | Sat. | 7:01 | 4:59 | 29 | 36 | 39 | snowed & sleeted last night |
| 26 | Sun. | 7:00 | 5:00 | 35 | 36 | 34 | |
| 27 | Mon. | 6:59 | 5:01 | 35 | 44 | 35 | |
| 28 | Tues. | 6:58 | 5:02 | 34 | 33 | 34 | |
| 29 | Wed. | 6:57 | 5:03 | 35 | 45 | 37 | |
| 30 | Thurs. | 6:56 | 5:04 | 38 | 35 | 36 | rained & drizzled all day |
| 31 | Fri. | 6:55 | 5:05 | 34 | 38 | 36 | |

February 1862

Veterans of the cold, cold Romney adventure found little relief in the new month. An officer wrote home on February 1: "The weather is cold—the ground being covered with an additional coat of snow, since last night." As the troops marched gratefully back toward Winchester from Romney on the 3rd, eight to ten inches of snow covered the ground. After wading streams choked with ice and "climbing the mountain road . . . so slick with ice, as to render an ascent almost impossible," they finally collapsed to sleep in the snow. In the relatively comfortable camps near Winchester, the troops enjoyed "bright, clear, cold" weather on the 12th, the sun reflecting off snowy fields "so dazzlingly, that it is painful to the eyes to look out upon it."[7]

The pervasive snow of this winter tormented a Georgian convalescing in a Winchester hospital in very different fashion than it hindered marching soldiers. Incapacitated for weeks with fever, as he "lay on my bed . . . and looked on the white-mantled hills that environed the town I remember distinctly how intensely my parched lips craved the cooling touch of the pure white snow"— but medical dogma of the era forbade cold water to feverish patients.[8]

A colonel in the sprawling camps around Manassas Junction and Centreville awoke at 3 a.m. on February 3 to find snow falling. It continued all day and into the night but did not accumulate much. On the 15th the colonel wrote in his diary, "snow storm all day."[9]

A Virginia artillerist camped near the tip of Virginia's Peninsula, a zone typically a good bit warmer than the rest of the state because of its proximity to the ocean, wrote cheerfully on February 10: "The winter . . . has been a mild one, with but little of snow."[10]

Horrible weather afflicted Jefferson Davis's inauguration on February 22. A citizen who attended said that the downpour made walking outdoors "as bad as I ever knew it."[11]

"Such a day!" a Confederate official wrote in his diary of February 22. "The heavens weep incessantly. Capitol Square is black with umbrellas." The rain lasted all day and all night. "The permanent government had its birth in storm, but it may yet flourish in sunshine," the official concluded optimistically if inaccurately.[12]

The Richmond weather station measured 6.7 inches of rain during the month, including two inches on the "rainy day" of the inauguration, and temperatures on that day ranging from 44 to 49 degrees. The precipitation included snow on February 3 and 6 and more rain on the 19th and 26th. Richmond temperatures for February averaged 36.7, 45.9, and 40.7 at 7 a.m., 2 p.m., and 9 p.m.[13]

Unfortunately, the thorough and legible records for Richmond weather end with this month. Apparently, Charles Meriwether had continued his weather ledger out of habit, even though his home no longer lay within the reach of the program's sponsors, rather like the famous case of the czar's postal system running for months after the Communist revolution. Not long after he stopped recording the weather, Meriwether put up for sale his "beautiful country seat called 'Westwood,' on the Brook Turnpike." The "country seat," which included sixty-three acres and three kinds of orchards, stood at a site that today is deep inside Richmond's city limits.[14]

February 1862

| Date | Day of Week | Rich-mond Sunrise | Rich-mond Sunset | D.C. Temperature | | | Other |
|---|---|---|---|---|---|---|---|
| | | | | 7 a.m. | 2 p.m. | 9 p.m. | |
| 1 | Sat. | 6:54 | 5:06 | 34 | 42 | 33 | |
| 2 | Sun. | 6:53 | 5:07 | 33 | 40 | 33 | |
| 3 | Mon. | 6:52 | 5:08 | 30 | 32 | 30 | snow all day |
| 4 | Tues. | 6:51 | 5:09 | 30 | 40 | 30 | |
| 5 | Wed. | 6:50 | 5:10 | 24 | 41 | 30 | |
| 6 | Thurs. | 6:49 | 5:11 | 34 | 40 | 40 | |
| 7 | Fri. | 6:48 | 5:12 | 41 | 46 | 36 | |
| 8 | Sat. | 6:47 | 5:13 | 36 | 37 | 36 | rainlike at 2 |
| 9 | Sun. | 6:46 | 5:14 | 33 | 42 | 30 | |
| 10 | Mon. | 6:45 | 5:15 | 17 | 35 | 26 | |
| 11 | Tues. | 6:44 | 5:16 | 30 | 38 | 35 | began snowing at 4 p.m. |
| 12 | Wed. | 6:43 | 5:17 | 34 | 50 | 46 | cloudy mackerel |
| 13 | Thurs. | 6:42 | 5:18 | 37 | 64 | 56 | |
| 14 | Fri. | 6:41 | 5:19 | 42 | 44 | 31 | sprinkling at 8 a.m. |
| 15 | Sat. | 6:40 | 5:20 | 31 | 37 | 27 | snow all day, 3 inches |
| 16 | Sun. | 6:39 | 5:21 | 36 | 40 | 32 | |
| 17 | Mon. | 6:37 | 5:23 | 30 | 34 | 37 | snow & sleet all day |
| 18 | Tues. | 6:36 | 5:24 | 35 | 47 | 37 | |
| 19 | Wed. | 6:35 | 5:25 | 35 | 38 | 38 | rain & sleet last night |
| 20 | Thurs. | 6:34 | 5:26 | 45 | 46 | 35 | |
| 21 | Fri. | 6:32 | 5:28 | 30 | 40 | 35 | |
| 22 | Sat. | 6:31 | 5:29 | 36 | 45 | 42 | fog & drizzle |
| 23 | Sun. | 6:30 | 5:30 | 40 | 46 | 57 | |
| 24 | Mon. | 6:29 | 5:31 | 48 | 46 | 31 | |
| 25 | Tues. | 6:28 | 5:32 | 22 | 41 | 28 | |
| 26 | Wed. | 6:27 | 5:33 | 31 | 50 | 42 | rain began 6 p.m. |
| 27 | Thurs. | 6:26 | 5:34 | 37 | 44 | 35 | |
| 28 | Fri. | 6:25 | 5:35 | 30 | 36 | 24 | high wind, 2 til sundown |

# March 1862

Men of a Tennessee regiment encamped on March 5 at Evansport, on the Potomac, found their location "low and muddy, the air damp and cold."[15]

On March 6 a Virginia artillerist posted near Yorktown described the ugly night just past: "Awakened this morning to find sleet, ice and snow. . . . Last night was one of the most miserable I ever spent, my limbs were nearly frozen, and to-day it is so very cold we have to keep wrapped up in our blankets all the time."[16]

The CSS *Virginia* steamed into Hampton Roads on March 8, a day that a Confederate naval officer described as "beautifully calm and clear." A French vessel in the harbor reported a "light breeze from the N.N.W." that morning and "very fine weather." March 9 dawned in this arena of intense naval interest with a "slight breeze from the east; very fine weather; light fog." The fog dispersed completely by 8 a.m., and the *Virginia* dueled with the USS *Monitor* under clear skies.[17]

As General Joseph E. Johnston planned one of the massive retrograde movements in which he specialized, the weather made his horribly congested logistical routes south from Manassas even less practicable for large-scale maneuvers. "The weather is still very bad," a soldier wrote home to his mother, "ground muddy and miry as it can be."[18]

During the mid-March retirement that took Confederate troops away from Manassas to the Rappahannock River line, the weather "remained cold and gloomy."[19] Soldiers slogging through mud faced soaking nights: "It rained every night just as soon as we would stop for the night."[20]

Soldiers near Mt. Jackson on the evening of March 20 "pitched our tents in a driving rain. . . . We have had some bad weather lately. I suppose it is the Vernal Equinox."[21]

General "Stonewall" Jackson fought the first battle of what would become his legendary Shenandoah Valley Campaign at Kernstown on March 23, a day called "muddy and disagreeable" by a young woman who lived not far away. She described both March 26 and 31 in the Valley as "bright and beautiful."

The USS *Monitor* survived a battering by the CSS *Virginia* in March 1862 in the famous first battle between iron-clad warships, but the weather accomplished what the Confederate navy could not: the *Monitor* foundered in a storm off the North Carolina coast on the last day of 1862. (*The Story of American Heroism: Thrilling Narratives of Personal Adventures during the Great Civil War as Told by the Medal Winners and Roll of Honor Men* [Chicago: The Werner Company, 1896], 673.)

The 29th in Charlottesville proved to be "a changeable day, sunshine, sleet, snow, Rain, thunder & lightning. A most shocking kind of day."[22]

On the lower Peninsula, rain fell on March 24. The next two days were "cloudy," then the 27th was "clear, cold" and the 28th a "fine day." Rain began on the night of March 29 and continued the next day.[23]

## March 1862

| Date | Day of Week | Rich-mond Sunrise | Rich-mond Sunset | D.C. Temperature | | | Other |
|---|---|---|---|---|---|---|---|
| | | | | 7 a.m. | 2 p.m. | 9 p.m. | |
| 1 | Sat. | 6:23 | 5:37 | 24 | 44 | 32 | |
| 2 | Sun. | 6:22 | 5:38 | 31 | 36 | 34 | 1" snow, drizzle at dark |
| 3 | Mon. | 6:21 | 5:39 | 34 | 37 | 38 | rained til 4 p.m., .40 |
| 4 | Tues. | 6:20 | 5:40 | 31 | 46 | 33 | rained last night, .63 |
| 5 | Wed. | 6:19 | 5:41 | 32 | 40 | 36 | |
| 6 | Thurs. | 6:17 | 5:43 | 30 | 44 | 36 | |
| 7 | Fri. | 6:16 | 5:44 | 24 | 40 | 30 | |
| 8 | Sat. | 6:15 | 5:45 | 31 | 49 | 35 | |
| 9 | Sun. | 6:14 | 5:46 | 29 | 65 | 42 | smoky at 7, heavy frost |
| 10 | Mon. | 6:12 | 5:48 | 46 | 56 | 36 | |
| 11 | Tues. | 6:11 | 5:49 | 36 | 57 | 40 | |
| 12 | Wed. | 6:10 | 5:50 | 33 | 67 | 55 | |
| 13 | Thurs. | 6:09 | 5:51 | 41 | 54 | 43 | began to drizzle at 2:30 |
| 14 | Fri. | 6:08 | 5:52 | 38 | 41 | 43 | |
| 15 | Sat. | 6:07 | 5:53 | 40 | 45 | 44 | rained all day, 1.62 |
| 16 | Sun. | 6:06 | 5:54 | 39 | 54 | 44 | last night .89 |
| 17 | Mon. | 6:04 | 5:56 | 34 | 45 | 41 | |
| 18 | Tues. | 6:03 | 5:57 | 36 | 49 | 39 | |
| 19 | Wed. | 6:02 | 5:58 | 37 | 49 | 46 | |
| 20 | Thurs. | 6:00 | 6:00 | 38 | 41 | 39 | Vernal Equinox, .37 |
| 21 | Fri. | 5:59 | 6:01 | 36 | 45 | 45 | yesterday .40 |
| 22 | Sat. | 5:58 | 6:02 | 39 | 53 | 42 | last night .14 |
| 23 | Sun. | 5:57 | 6:03 | 40 | 52 | 43 | |
| 24 | Mon. | 5:55 | 6:05 | 39 | 50 | 41 | |
| 25 | Tues. | 5:54 | 6:06 | 33 | 48 | 37 | white frost |
| 26 | Wed. | 5:53 | 6:07 | 29 | 51 | 40 | |
| 27 | Thurs. | 5:52 | 6:08 | 32 | 54 | 39 | white frost slight |
| 28 | Fri. | 5:51 | 6:09 | 35 | 62 | 54 | |
| 29 | Sat. | 5:50 | 6:10 | 34 | 32 | 33 | |
| 30 | Sun. | 5:49 | 6:11 | 33 | 42 | 34 | snow 1.5 |
| 31 | Mon. | 5:48 | 6:12 | 39 | 62 | 47 | last night rain .14 |

April 1862

The log of the gunboat USS *Wachusett,* stationed on the lower James River, reported "more or less rain each day" on April 7–9 near the tip of the Peninsula.[24]

After marching through Orange Court House toward Fredericksburg on the night of April 6, the men of the 11th Georgia halted at about 10 a.m. on the 7th. They did not move for twenty-four hours, "snow, rain and sleet falling the whole while in rapid alternations, and presenting a painful exhibition of the fierce, relentless anger of the savage, ill-tempered and intractable elements."[25]

A woman living in Fredericksburg expressed regret in her diary on April 8 over the sufferings of Confederate troops marching past: "It has been raining hard all day, and they were wet to the skin; but they all looked bright and cheerful."[26]

Citizens of Front Royal awoke on the morning of April 9 to find "snow eight or ten inches deep on the ground," and a scene of dazzling beauty as the sun glinted off ice haloes around peach blossoms and violets. More bright sun on the 10th produced another fairy-tale scene.[27]

According to contemporary diaries, the weather on Virginia's Peninsula remained "clear & pleasant," or "fine," or even "delightful" from April 11 to April 18, including the day of battle along the Warwick River on the 16th.[28]

On April 18, local cavalrymen forced to abandon Fredericksburg to the enemy for the first time left their home grounds under a bright sky. As they looked back across the Rappahannock River they could see enemy regiments "marching . . . across the fields of the Chatham estate with their muskets glittering brightly in the morning sun."[29]

From a temperate zone near Suffolk, a Virginia artillerist waxed almost giddy about the weather in his diary on April 23. His note about the human tendency to report storms instead of good weather surely applies to most diaries and contemporary writings: "What a beautiful day! How prone we are to note only the *stormy* days, whilst the days of beauty seem to be taken as be-longing to us by right. . . . All nature shines resplendent in the soft beauties of a spring-day morn; the noble oaks and stately elms are budding forth . . . and the earth is with verdure clad."[30]

April closed wetly in Fredericksburg. After a "cloudy and damp" morning on the 29th, with a "slight rain," "a drizzly rain" fell all morning on the 30th.[31]

April 1862

| Date | Day of Week | Rich-mond Sunrise | Rich-mond Sunset | D.C. Temperature | | | Other |
|---|---|---|---|---|---|---|---|
| | | | | 7 a.m. | 2 p.m. | 9 p.m. | |
| 1 | Tues. | 5:46 | 6:14 | 39 | 65 | 50 | |
| 2 | Wed. | 5:45 | 6:15 | 44 | 52 | 47 | |
| 3 | Thurs. | 5:44 | 6:16 | 36 | 65 | 52 | |
| 4 | Fri. | 5:43 | 6:17 | 39 | 66 | 63 | |
| 5 | Sat. | 5:41 | 6:19 | 47 | 52 | 43 | drizzly, .14 |
| 6 | Sun. | 5:40 | 6:20 | 37 | 63 | 50 | very heavy frost |
| 7 | Mon. | 5:39 | 6:21 | 44 | 45 | 36 | snow & rain from 1 p.m., 2.00 |
| 8 | Tues. | 5:38 | 6:22 | 34 | 37 | 34 | rained & sleeted, 1.55 |
| 9 | Wed. | 5:36 | 6:24 | 32 | 37 | 35 | snowed last night 2.00 |
| 10 | Thurs. | 5:35 | 6:25 | 31 | 46 | 41 | |
| 11 | Fri. | 5:34 | 6:26 | 36 | 53 | 40 | |
| 12 | Sat. | 5:33 | 6:27 | 34 | 55 | 44 | |
| 13 | Sun. | 5:32 | 6:28 | 40 | 62 | 51 | |
| 14 | Mon. | 5:31 | 6:29 | 48 | 68 | 54 | |
| 15 | Tues. | 5:30 | 6:30 | 47 | 70 | 60 | |
| 16 | Wed. | 5:29 | 6:31 | 56 | 73 | 60 | |
| 17 | Thurs. | 5:28 | 6:32 | 57 | 82 | 67 | |
| 18 | Fri. | 5:26 | 6:34 | 62 | 84 | 66 | rain .85 |
| 19 | Sat. | 5:25 | 6:35 | 62 | 62 | 57 | peach & cherry blos-soms beg. to develop |
| 20 | Sun. | 5:24 | 6:36 | 49 | 50 | 48 | Easter Sunday, showery |
| 21 | Mon. | 5:23 | 6:37 | 42 | 46 | 50 | rained all day .45 |
| 22 | Tues. | 5:22 | 6:38 | 48 | 58 | 48 | |
| 23 | Wed. | 5:21 | 6:39 | 48 | 45 | 48 | |
| 24 | Thurs. | 5:20 | 6:40 | 42 | 56 | 46 | |
| 25 | Fri. | 5:19 | 6:41 | 41 | 47 | 46 | |
| 26 | Sat. | 5:18 | 6:42 | 45 | 60 | 47 | |
| 27 | Sun. | 5:17 | 6:43 | 46 | 60 | 39 | white frost |
| 28 | Mon. | 5:16 | 6:44 | 43 | 62 | 52 | |
| 29 | Tues. | 5:15 | 6:45 | 52 | 61 | 52 | sprinkled rain |
| 30 | Wed. | 5:14 | 6:46 | 52 | 62 | 56 | heavy dew, drizzle from 3 |

## May 1862

Southerners who fought at Williamsburg remembered the pitiless rain that pelted that field steadily on May 5. A member of the 5th South Carolina described the night before the battle: "All night long it rained—poured down in torrents." On May 6, "again all day and night it poured down rain."[32]

A New Yorker fighting the Carolinians got just as wet. His regiment marched through "a heavy rain storm" during the night of May 4–5 and completely lost its course. "Muddy roads" kept ammunition resupply from reaching the front, but the men advanced through "deep mud and drenching rain." A bandsman assigned to bring in wounded soldiers could not do much good. "We could hear wounded men between the picket lines lying in the cold rain begging to be carried off the field," he wrote, "but an attempt to do so would bring a shower of balls."[33]

Confederates marshaled outside Richmond, after retreating up the Peninsula through bottomless mud, in weather "clear and hot" on May 12 and 13, and rainy on the evening of May 14 and on into May 16. The sky then turned sunny and "cloudy and hot" on the 17th, and "clear and hot" on the 18th and 20th.[34]

"A disagreeable, rainy, misty morning" opened May in the Shenandoah Valley, where famous events would unfold during this month. The 4th was "so pleasant" in Front Royal, the 6th brought "bright May sunshine," and still on the 9th the Valley weather remained "so beautiful," in the words of a local woman. Invading Northern troops swarmed through Front Royal, where it was "raining very hard," on May 14; the next day continued "cloudy and rainy." A nineteen-year-old woman in Front Royal found the afternoon of May 23, just before "Stonewall" Jackson roared into town and liberated it from occupation, to be "oppressively warm."[35]

On the evening of Jackson's great victory at Winchester on May 25, a Fredericksburg diarist reveled in "a beautiful day"; "there could not be a more favourable one for a march."[36]

Jackson's advanced troops fell back from Halltown on May 30 "in a terrible rain storm about four p.m."[37]

After the rain-drenched Battle of Williamsburg on May 5, 1862, Confederates retreating toward Richmond and Federals pursuing them both struggled along roads so choked with mud that mules sank entirely from view in the gumbo. (Robert Underwood Johnson and C. C. Buel, eds., *Battles and Leaders of the Civil War: Being for the Most Part Contributions by Union and Confederate Officers,* 4 vols. [New York: The Century Co., 1887–88], 2:192.)

North Carolinians fought the Battle of Slash Church (also known as Hanover Court House) north of Richmond on May 27, a day during which skies cleared from rain the night before. By the end of the day, the deck log of a Federal gunboat near Petersburg called the weather "fine."[38]

A soldier who fought at Seven Pines on May 31 wrote: "The preceding night had been one of great storm. The streams were flooded." In the "murky morning" of the next day, musket smoke only slowly "curled up through the damp trees."[39]

In a camp near Richmond, the "very severe thunderstorm" on the night of May 30 brought "wind and a deluge of rain. Many tents were blown down and the camp flooded. The lightning was very sharp and the clouds hung over us for several hours."[40] Another Confederate described "the worst thunder storms I have ever known. Storm after storm from early in the afternoon until midnight, and when we got to the creek just below Richmond it was like a river."[41]

The violent storm of May 30 killed at least one Federal soldier, a member of the 44th New York: "During the afternoon occurred a terrific thunder storm, during which lightning struck the tent of Quartermaster Sergeant [Henry C.] Howlett and Sergeant Major [John B.] Web[b]er, instantly killing the former, rendering the latter insensible and igniting and exploding a box of cartridges."[42]

May 1862

| Date | Day of Week | Rich-mond Sunrise | Rich-mond Sunset | D.C. Temperature | | | Other |
|------|------|------|------|------|------|------|------|
| | | | | 7 a.m. | 2 p.m. | 9 p.m. | |
| 1 | Thurs. | 5:13 | 6:47 | 51 | — | — | 3-reading avg. = 54.3 |
| 2 | Fri. | 5:12 | 6:48 | 56 | 6- | — | avg. = 57.3 |
| 3 | Sat. | 5:11 | 6:49 | 55 | — | — | avg. = 65.2 |
| 4 | Sun. | 5:10 | 6:50 | 58 | 7- | — | avg. = 62.7 |
| 5 | Mon. | 5:09 | 6:51 | 57 | — | — | avg. = 65.3 |
| 6 | Tues. | 5:08 | 6:52 | 52 | 50 | 52 | began drizzly |
| 7 | Wed. | 5:07 | 6:53 | 49 | 72 | 60 | |
| 8 | Thurs. | 5:06 | 6:54 | 52 | 68 | 54 | sprinkle at 6 p.m. |
| 9 | Fri. | 5:05 | 6:55 | 52 | 80 | 64 | |
| 10 | Sat. | 5:04 | 6:56 | 65 | 86 | 68 | |
| 11 | Sun. | 5:03 | 6:57 | 59 | 68 | 56 | dim stars |
| 12 | Mon. | 5:02 | 6:58 | 49 | 76 | 60 | smoky & dustlike all day |
| 13 | Tues. | 5:01 | 6:59 | 60 | — | 73 | |
| 14 | Wed. | 5:00 | 7:00 | 64 | 56 | 57 | rained slightly all day |
| 15 | Thurs. | 4:59 | 7:01 | 54 | 60 | 57 | yesterday & today .49 |
| 16 | Fri. | 4:58 | 7:02 | 62 | 77 | 64 | |
| 17 | Sat. | 4:57 | 7:03 | 64 | 80 | 62 | |
| 18 | Sun. | 4:57 | 7:03 | 66 | 83 | 66 | occasional lightning 9 p.m. |
| 19 | Mon. | 4:56 | 7:04 | 68 | 60 | 58 | |
| 20 | Tues. | 4:55 | 7:05 | 68 | 74 | 64 | rainlike at 10 p.m. |
| 21 | Wed. | 4:55 | 7:05 | 60 | 84 | 75 | falling mist |
| 22 | Thurs. | 4:54 | 7:06 | 78 | 86 | 68 | |
| 23 | Fri. | 4:53 | 7:07 | 62 | 86 | 67 | refreshing shower at 5 p.m., 30 min. |
| 24 | Sat. | 4:53 | 7:07 | 60 | 56 | 49 | .35 yesterday & today |
| 25 | Sun. | 4:52 | 7:08 | 54 | 70 | 56 | heavy dew |
| 26 | Mon. | 4:51 | 7:09 | 63 | 75 | 63 | |
| 27 | Tues. | 4:51 | 7:09 | 59 | 77 | 63 | drizzly |
| 28 | Wed. | 4:50 | 7:10 | 71 | 75 | 54 | |
| 29 | Thurs. | 4:49 | 7:11 | 62 | 76 | 62 | |
| 30 | Fri. | 4:49 | 7:11 | 69 | 85 | 63 | showers in p.m. |
| 31 | Sa | 4:48 | 7:12 | 62 | 70 | 60 | drizzly all day |

June 1862

General Samuel P. Heintzelman's journal called June 1 at Savage's Station "warm." A little rain early the following morning produced a rainbow. That night the general recorded "heavy rain and a thunderstorm," followed by a hot morning: "Mercury, at one time, in my tent, 94." Each day through June 7, Heintzelman wrote of clouds and rain.[43]

Another lightning bolt killed a second New York soldier on the morning of June 3, just four days after Howlett's death. "A terrific discharge of the electric fluid" flashed down from the skies and sizzled through the tents of Battery E, 1st New York Light Artillery. It stunned or burned twenty men, knocking several of them unconscious, and instantly killed Corporal James Bryant. *Harper's Weekly* published the story, illustrated by an artist's depiction of the event, four weeks later.[44]

A Fredericksburg woman wrote in her diary on June 4: "All the Yankee bridges were washed away today by the freshet." The same woman headed south on June 25 but ran into the aftermath of another downpour: "We were out in a dreadful storm, and got wet through and through. . . . All the bridges have been destroyed, and we can only cross by fording."[45]

Jackson fought his battles at Cross Keys and Port Republic on June 8 and 9, concluding the Shenandoah Valley Campaign that made him world famous, under bright skies and a pleasant late-spring atmosphere. A Harrisonburg diarist described the next three days as "Cloudy and rain" (June 10), "Clear and very cool" (June 11), and "Clear and warm" (June 12).[46]

Captain Beckwith West of the 48th Virginia, having been captured at Front Royal, reached Washington at noon on June 10 aboard a steamboat as a prisoner. The prisoners marched to Old Capitol Prison "very wet . . . on account of the rain which commenced before we left the steamboat."[47]

Back near Richmond, General Heintzelman's journal noted rain most of the night of June 9 and into the following morning. On the 15th a powerful front moved through, with thunder and lightning and rain, "and the air cooled greatly," dropping from 93 to 66 degrees. Cool weather held on through

A lightning strike killed New Yorker James Bryant on June 3, 1862. *Harper's Weekly* reported the episode a few weeks later and illustrated the story with an artist's conception of the disastrous moment. (*Harper's Weekly*, July 5, 1862, 426.)

June 19, then warmed steadily. Heintzelman's thermometer measured 93 degrees on June 22.[48]

A Maine soldier detailed to guard some cattle overnight on June 11 near Richmond marveled in his diary: "During the night an almost total eclipse of the moon took place, the first I ever saw."[49]

As the momentous battles for Richmond approached, "a most furious & tremendous thunder storm & a very hard rain" fell on King and Queen County on the night of June 23. The downpour continued all night long in Fredericksburg and washed away the Yankee pontoons in the river. June 25 in Fredericksburg brought "cool delightful" weather.[50]

A thermometer aboard the USS *Galena*, anchored in the James River, recorded 84 degrees at noon on June 23 and 86 at noon on the 24th. The deck log noted a clear day on the 25th, with temperatures in the low 70s.[51]

Late on June 26, Richmond citizens who gathered on the city's high points to look toward the battlefields to northeastward felt "a gentle breeze rising" as darkness fell, revealing "the glimmer of the stars."[52]

After "a bright clear day" with "light breezes" on June 27, during which the Battle of Gaines's Mill raged, it "rained heavily" on the night of June 28 in some areas around Richmond, "although apparently in a limited extent of the country." Sailors on nearby rivers recorded temperatures of 85 degrees at

Union General Samuel P. Heintzelman frequently reported weather observations in his diary. (Benson J. Lossing, *Pictorial History of the Civil War in the United States of America,* 3 vols. [Philadelphia: G. W. Childs, 1866–68], 2:417.)

1 p.m. and 80 degrees at 4 p.m. on the 28th. On the night of June 29, "A heavy thunder storm" lashed the battlefields. Near White House on the Pamunkey River, the log of the USS *Sebago* reported that the storm lasted for nearly six hours, with varying intensity, and that during two hours "the rain fell in torrents."[53]

Soldiers originating from Virginia's western mountains and valleys, or who originally had come from elsewhere but had been fighting in that region for some months, found the atmosphere in the Richmond swamps debilitating. "The day was warm," one of them wrote of June 29, "and the roads dusty and the march very fatiguing."[54]

June 1862

| Date | Day of Week | Rich-mond Sunrise | Rich-mond Sunset | D.C. Temperature | | | Other |
|---|---|---|---|---|---|---|---|
| | | | | 7 a.m. | 2 p.m. | 9 p.m. | |
| 1 | Sun. | 4:48 | 7:12 | 63 | 78 | 71 | drizzle .40, sharp thunder at 8 p.m. |
| 2 | Mon. | 4:47 | 7:13 | 72 | 84 | 72 | |
| 3 | Tues. | 4:47 | 7:13 | 71 | 92 | 76 | .15 last night |
| 4 | Wed. | 4:46 | 7:14 | 64 | 82 | 66 | last night & today 1.93 |
| 5 | Thurs. | 4:46 | 7:14 | 65 | 71 | 62 | drizzled all day .12 |
| 6 | Fri. | 4:45 | 7:15 | 66 | 70 | 62 | first firefly seen |
| 7 | Sat. | 4:45 | 7:15 | 69 | 82 | 60 | |
| 8 | Sun. | 4:44 | 7:16 | 63 | 72 | 57 | last night .15 |
| 9 | Mon. | 4:44 | 7:16 | 60 | 60 | 56 | |
| 10 | Tues. | 4:44 | 7:16 | 57 | 56 | 56 | |
| 11 | Wed. | 4:44 | 7:16 | 57 | 71 | 64 | |
| 12 | Thurs. | 4:43 | 7:17 | 70 | 88 | 70 | |
| 13 | Fri. | 4:43 | 7:17 | 72 | 91 | 70 | |
| 14 | Sat. | 4:43 | 7:17 | 79 | 91 | 78 | |
| 15 | Sun. | 4:43 | 7:17 | 78 | 71 | 63 | last night .11 |
| 16 | Mon. | 4:42 | 7:18 | 54 | 68 | 54 | |
| 17 | Tues. | 4:42 | 7:18 | 58 | 74 | 63 | |
| 18 | Wed. | 4:42 | 7:18 | 68 | 83 | 70 | |
| 19 | Thurs. | 4:42 | 7:18 | 70 | 80 | 63 | showery all day |
| 20 | Fri. | 4:42 | 7:18 | 65 | 74 | 61 | |
| 21 | Sat. | 4:42 | 7:18 | 68 | 84 | 66 | Summer Solstice |
| 22 | Sun. | 4:42 | 7:18 | 74 | 82 | 71 | |
| 23 | Mon. | 4:42 | 7:18 | 71 | 81 | 68 | smoky sun at 7 |
| 24 | Tues. | 4:42 | 7:18 | 67 | 80 | 70 | |
| 25 | Wed. | 4:42 | 7:18 | 65 | 76 | 70 | |
| 26 | Thurs. | 4:42 | 7:18 | 64 | 84 | 74 | |
| 27 | Fri. | 4:43 | 7:17 | 70 | 88 | 72 | |
| 28 | Sat. | 4:43 | 7:17 | 70 | 86 | 75 | |
| 29 | Sun. | 4:43 | 7:17 | 74 | 83 | 76 | shower at 4:30 p.m. |
| 30 | Mon. | 4:43 | 7:17 | 72 | 84 | 74 | rained last night .22 |

## July 1862

"Towards morning" on July 2 "it rained very hard for about three hours" in the vicinity of blood-soaked Malvern Hill. Another storm hit Richmond on the 11th: "raining very hard all day. . . . The sombre clouds of a gloomy day often cast an equal gloom over our spirits."[55]

The bad weather affected pursuit of General George B. McClellan's Federals after the Seven Days Battles around Richmond. "We rose . . . in a heavy rain, which continued throughout the day," a Virginia soldier wrote of July 2. "Our long fatiguing march and the fact that we were wet to the skin from the rain . . . prevented anything like rapid pursuit [during] a toilsome, disagreeable, all-day march."[56]

The "terrific storm" moved on eastward and later on July 2 unloaded downpours and hail on Hampton Roads, accompanied by lightning and thunder.[57]

Temperature readings from the deck log of the USS *Galena* provide a good picture of the weather just below Richmond in early July: 74 degrees at 11 a.m. on July 1; 60 degrees all afternoon on the 2nd; mid-60s on the 3rd; mid-90s by the 9th; and 102 degrees at 11 a.m. on the 10th.[58]

In Fredericksburg, July 8 was "very hot," and the 9th was "hotter than ever, and still dry." Relief came on the 10th in the form of "mist and cloud with a slight breeze, the change . . . delightful." The next two days also remained pleasant.[59]

On July 15, "at dark a heavy thunder storm" lashed the Federal camps around Harrison's Landing, abruptly lowering the temperature from 96 to 74 degrees. Temperatures recorded by Federal gunboat logs went up into the mid-90s on the 16th, and to 95 at 1 p.m. on the 17th.[60]

Near Petersburg, July 29 was a "fine day" and July 31 was "rainy."[61]

The retreat from Malvern Hill. (Alfred H. Guernsey and Henry M. Alden, eds., *Harper's Pictorial History of the Civil War* [Chicago: Star Publishing Company, 1894], 377.)

The deck log of the USS *Galena* for June and July 1862 recorded useful details about the weather on the James River just below Richmond. (Francis Trevelyan Miller, ed., *The Photographic History of the Civil War in Ten Volumes*, 10 vols. [New York: The Review of Reviews Co., 1911], 6:[137].)

## July 1862

| Date | Day of Week | Rich-mond Sunrise | Rich-mond Sunset | D.C. Temperature | | | Other |
|---|---|---|---|---|---|---|---|
| | | | | 7 a.m. | 2 p.m. | 9 p.m. | |
| 1 | Tues. | 4:43 | 7:17 | 66 | 77 | 65 | |
| 2 | Wed. | 4:44 | 7:16 | 62 | 76 | 57 | last night 1.60 |
| 3 | Thurs. | 4:44 | 7:16 | 62 | 70 | 66 | |
| 4 | Fri. | 4:44 | 7:16 | 62 | 78 | 64 | last night .22 |
| 5 | Sat. | 4:45 | 7:15 | 68 | 86 | 77 | |
| 6 | Sun. | 4:45 | 7:15 | 74 | 91 | 82 | |
| 7 | Mon. | 4:45 | 7:15 | 80 | 95 | 74 | heavy rain, hail 6:30–7 |
| 8 | Tues. | 4:46 | 7:14 | 75 | 92 | 83 | |
| 9 | Wed. | 4:46 | 7:14 | 81 | 94 | 84 | |
| 10 | Thurs. | 4:47 | 7:13 | 74 | 73 | 71 | windy last night |
| 11 | Fri. | 4:47 | 7:13 | 67 | 81 | 68 | |
| 12 | Sat. | 4:48 | 7:12 | 66 | 82 | 70 | smoky & hazy all day |
| 13 | Sun. | 4:48 | 7:12 | 70 | 88 | 73 | |
| 14 | Mon. | 4:49 | 7:11 | 72 | 88 | 78 | |
| 15 | Tues. | 4:50 | 7:10 | 80 | 76 | 77 | high wind from W |
| 16 | Wed. | 4:50 | 7:10 | 78 | 72 | 70 | rain to 5 p.m., heavy gust 2–3:30 |
| 17 | Thurs. | 4:51 | 7:09 | 74 | 85 | 76 | |
| 18 | Fri. | 4:52 | 7:08 | 66 | 70 | 76 | last night 1.02, drizzled all day |
| 19 | Sat. | 4:52 | 7:08 | 67 | 74 | 71 | |
| 20 | Sun. | 4:53 | 7:07 | 70 | 85 | 74 | |
| 21 | Mon. | 4:54 | 7:06 | 74 | 84 | 72 | |
| 22 | Tues. | 4:55 | 7:05 | 69 | 80 | 70 | |
| 23 | Wed. | 4:55 | 7:05 | 72 | 82 | 73 | |
| 24 | Thurs. | 4:56 | 7:04 | 74 | 90 | 73 | |
| 25 | Fri. | 4:57 | 7:03 | 74 | 87 | 74 | |
| 26 | Sat. | 4:57 | 7:03 | 71 | 85 | 71 | began very hard rain, 9 p.m. |
| 27 | Sun. | 4:58 | 7:02 | 80 | 83 | 74 | |
| 28 | Mon. | 4:59 | 7:01 | 73 | 88 | 77 | |
| 29 | Tues. | 4:59 | 7:01 | 77 | 84 | 77 | 3 p.m. thunder |
| 30 | Wed. | 5:00 | 7:00 | 73 | 90 | 78 | |
| 31 | Thurs. | 5:01 | 6:59 | 75 | 88 | 74 | heavy shower 4–5 p.m. |

## August 1862

"Pleasant" and "fine" weather opened August near Richmond on the first two days of the month, but August 3 turned "wet & disagreeable." Clear weather prevailed then until a rain squall came through on the evening of the 10th, and continued into the night.[62]

General "Stonewall" Jackson fought his last independent action, at Cedar Mountain on August 9, under broiling temperatures—the highest recorded (98) on the day of battle for any major action in the Virginia theater.[63]

A North Carolinian wrote from camp southeast of Richmond on August 11: "It rained yesterday morning, and the ground was wet. . . . It's been awfully hot here today. I believe it is warmer here than at home."[64]

The raid by General J. E. B. Stuart and his horsemen against General John Pope's supply lines at Catlett's Station on August 22 reached its climax during a torrential downpour. According to a member of the 1st Virginia Cavalry, "a terrific thunderstorm was raging, and the rain falling in torrents, and the darkness so intense that the whitest house could not be seen immediately in front of you."[65]

Rain fell "so steadily" in Botetourt County on August 24 that a pious family could not get out to church.[66]

"Tuesday, the 26th, the sun rose beautifully and the entire day was perfect" as Lee's army approached the plains of Manassas for the war's second great battle there.[67]

The armies fought during August 27–30 in weather that a diarist with Lee's army called "very fine" on the 27th, "fine" on the 28th, "a fine day, shower in the evening" on the 29th, and "a fine warm day, but rain in the p.m." on the 30th.[68]

Hot, dry weather made possible a dusty ruse near Manassas on August 29. To deceive Federals into thinking Confederate reinforcements had arrived, J. E. B. Stuart "caused his cavalry to get branches . . . tying them to their

On August 29, 1862, Jeb Stuart's Confederate cavalry dragged branches through deep dust to convince the enemy that reinforcements had arrived. (*The Story of American Heroism: Thrilling Narratives of Personal Adventures during the Great Civil War as Told by the Medal Winners and Roll of Honor Men* [Chicago: The Werner Company, 1896], 336.)

halter straps, to gallop backwards and forwards, dragging them on a dusty road . . . where the clouds of dust rose up in sight of the enemy."[69]

Rain poured on the Manassas battlefield the day after Lee's victory. "August 31st, 1862, we rested as much as it was possible for us to do, considering the weather. It was raining hard and there were no living Yankees in sight, but plenty of dead ones."[70]

August 1862

| Date | Day of Week | Rich-mond Sunrise | Rich-mond Sunset | D.C. Temperature | | | Other |
|---|---|---|---|---|---|---|---|
| | | | | 7 a.m. | 2 p.m. | 9 p.m. | |
| 1 | Fri. | 5:02 | 6:58 | 73 | 89 | 76 | |
| 2 | Sat. | 5:03 | 6:57 | 70 | 88 | 78 | |
| 3 | Sun. | 5:04 | 6:56 | 78 | 86 | 80 | 1.46, heavy shower at 2 p.m. |
| 4 | Mon. | 5:05 | 6:55 | 76 | 90 | 81 | |
| 5 | Tues. | 5:06 | 6:54 | 79 | 94 | 83 | |
| 6 | Wed. | 5:07 | 6:53 | 80 | 89 | 80 | |
| 7 | Thurs. | 5:08 | 6:52 | 76 | 92 | 83 | |
| 8 | Fri. | 5:09 | 6:51 | 80 | 96 | 86 | |
| 9 | Sat. | 5:10 | 6:50 | 84 | 98 | 84 | |
| 10 | Sun. | 5:11 | 6:49 | 86 | 94 | 84 | sprinkly at 9 p.m. |
| 11 | Mon. | 5:12 | 6:48 | 80 | 95 | 83 | |
| 12 | Tues. | 5:13 | 6:47 | 79 | 90 | 75 | shower, thunder at 3 p.m. |
| 13 | Wed. | 5:14 | 6:46 | 72 | 85 | 71 | slight shower last night |
| 14 | Thurs. | 5:15 | 6:45 | 66 | 86 | 78 | |
| 15 | Fri. | 5:16 | 6:44 | 74 | 84 | 66 | |
| 16 | Sat. | 5:17 | 6:43 | 63 | 76 | 62 | |
| 17 | Sun. | 5:18 | 6:42 | 59 | 76 | 66 | |
| 18 | Mon. | 5:19 | 6:41 | 64 | 78 | 64 | |
| 19 | Tues. | 5:20 | 6:40 | 66 | 82 | 65 | |
| 20 | Wed. | 5:21 | 6:39 | 65 | 84 | 76 | |
| 21 | Thurs. | 5:22 | 6:38 | 70 | 86 | 79 | |
| 22 | Fri. | 5:23 | 6:37 | 74 | 88 | 80 | drizzly at 8 p.m. |
| 23 | Sat. | 5:24 | 6:36 | 77 | 89 | 76 | |
| 24 | Sun. | 5:25 | 6:35 | 69 | 74 | 64 | |
| 25 | Mon. | 5:27 | 6:33 | 59 | 78 | 68 | |
| 26 | Tues. | 5:28 | 6:32 | 64 | 88 | 77 | |
| 27 | Wed. | 5:29 | 6:31 | 70 | 90 | 78 | |
| 28 | Thurs. | 5:30 | 6:30 | 74 | 80 | 72 | slight shower last night |
| 29 | Fri. | 5:31 | 6:29 | 68 | 88 | 76 | |
| 30 | Sat. | 5:32 | 6:28 | 66 | 79 | 71 | |
| 31 | Sun. | 5:33 | 6:27 | 65 | 71 | 73 | rained & hailed at night |

September 1862

Everyone near the Chantilly battlefield on September 1 remarked upon the merciless downpour that interrupted the action and led to several memorable anecdotes based on the weather. A soldier in the 13th South Carolina recalled the torrents: "There was a cloud over us [and] I never experienced [such] thunder and lightning, it was almost incessant." Union General Philip Kearny, a seasoned and much-admired officer destined for promotion, was killed when he rode into a Confederate regiment in the confusion resulting from the storm.[71]

Confederates marching from the environs of Richmond across northern Virginia toward the Potomac River in early September straggled terribly. "It has been an awfully hard march," one of them wrote on September 5. "Two men died in one day from sun stroke. The weather is not so warm now as some days ago." In fact, the nights had begun to turn cool.[72]

Troops crossing the Potomac into Maryland after dark on September 7 enjoyed "a beautiful night; the crystalline waters of the lovely Potomac danced and sparkled in the soft moonbeams."[73]

On September 10 in Arlington, "a heavy wind storm" blew through, but brought "scarcely rain enough to lay the dust." The next morning, showers fell there until noon.[74]

On the shoulder of the Blue Ridge not far south of the Potomac River, September 14 was "a gloomy, sad day . . . the cloudy veil shrouding the sky." The 19th dawned there as "a charming morning—bright, sunshine and a fine autumn breeze."[75]

Death of General Philip Kearny, September 1, 1862. (Robert Underwood Johnson and
C. C. Buel, eds., *Battles and Leaders of the Civil War: Being for the Most Part Contributions by Union and Confederate Officers,* 4 vols. [New York: The Century Co., 1887–88],
2:537.)

September 1862

| Date | Day of Week | Rich-mond Sunrise | Rich-mond Sunset | D.C. Temperature | | | Other |
|------|-------------|-------------------|------------------|------|------|------|-------|
| | | | | 7 a.m. | 2 p.m. | 9 p.m. | |
| 1 | Mon. | 5:34 | 6:26 | 74 | 83 | 67 | began rain at 6 p.m., .86 |
| 2 | Tues. | 5:35 | 6:25 | 60 | 68 | 61 | |
| 3 | Wed. | 5:36 | 6:24 | 60 | 72 | 57 | |
| 4 | Thurs. | 5:38 | 6:22 | 57 | 77 | 62 | |
| 5 | Fri. | 5:39 | 6:21 | 57 | 80 | 71 | |
| 6 | Sat. | 5:40 | 6:20 | 64 | 84 | 78 | |
| 7 | Sun. | 5:41 | 6:19 | 70 | 85 | 75 | |
| 8 | Mon. | 5:42 | 6:18 | 67 | 88 | 78 | |
| 9 | Tues. | 5:44 | 6:16 | 71 | 81 | 76 | rainlike at 9 p.m. |
| 10 | Wed. | 5:45 | 6:15 | 69 | 8- | 73 | sprinkled at 5:30 |
| 11 | Thurs. | 5:46 | 6:14 | 73 | 85 | 75 | showery all day |
| 12 | Fri. | 5:47 | 6:13 | 72 | 84 | 78 | .47 last night |
| 13 | Sat. | 5:48 | 6:12 | 68 | 77 | 63 | |
| 14 | Sun. | 5:49 | 6:11 | 61 | 75 | 61 | |
| 15 | Mon. | 5:50 | 6:10 | 69 | 84 | 70 | |
| 16 | Tues. | 5:51 | 6:09 | 75 | 74 | 68 | air seemed filled w/mist |
| 17 | Wed. | 5:53 | 6:07 | 69 | 76 | 70 | last night .22 |
| 18 | Thurs. | 5:54 | 6:06 | 73 | 84 | 76 | last night .07 |
| 19 | Fri. | 5:55 | 6:05 | 68 | 83 | 71 | |
| 20 | Sat. | 5:56 | 6:04 | 64 | 75 | 66 | sprinkled |
| 21 | Sun. | 5:58 | 6:02 | 62 | 78 | 66 | sky very red at sunset |
| 22 | Mon. | 5:59 | 6:01 | 57 | 79 | 63 | |
| 23 | Tues. | 6:00 | 6:00 | 56 | 78 | 60 | Autumnal Equinox |
| 24 | Wed. | 6:01 | 5:59 | 64 | 76 | 58 | |
| 25 | Thurs. | 6:03 | 5:57 | 66 | 68 | 64 | |
| 26 | Fri. | 6:04 | 5:56 | 53 | 69 | 52 | |
| 27 | Sat. | 6:05 | 5:55 | 50 | 72 | 54 | |
| 28 | Sun. | 6:07 | 5:53 | 64 | 72 | 67 | slight shower last night |
| 29 | Mon. | 6:08 | 5:52 | 64 | 80 | 70 | |
| 30 | Tues. | 6:09 | 5:51 | 59 | 82 | 76 | |

## October 1862

Richmonders witnessed "a splendid aurora borealis" on the night of October 3.[76]

Confederates encamped near Bunker Hill on October 8 were "pining for rain and oppressed by the heat [and by] clouds of dust. . . . It is as hot as midsummer here, and every breath of air that stirs, instead of bringing coolness, bears as great a load of dust as ever a wind that blew over the desert of Sahara."[77]

In Fairfax County, a woman wrote on October 4: "The weather is as hot now as it has been any time during the summer. And to-day the scorching wind is thick with dust." On the 13th it rained "nearly the whole day." Rain "poured" on the 26th and continued until noon on the 27th.[78]

Troops at Chaffin's Bluff on October 12 wore overcoats after dinner to keep warm. One of them mused ruefully about the joys of sitting around a home hearth on autumn evenings of this kind, and wrote of working on winter quarters.[79]

"It rained very hard the first part of the night," a Virginian in camp near Bunker Hill wrote on October 16, "but, after midnight, cleared off."[80]

Dawn of October 21 came with "the biggest Frost" in King and Queen County, where "the wind has been blowing hard all day, like March weather." Steady rain fell there on October 26.[81]

October 1862

| Date | Day of Week | Rich-mond Sunrise | Rich-mond Sunset | D.C. Temperature | | | Other |
|------|-------------|-------------------|------------------|-----------------|---------|---------|-------|
| | | | | 7 a.m. | 2 p.m. | 9 p.m. | |
| 1 | Wed. | 6:10 | 5:50 | 69 | 82 | 71 | sprinkly at 7 |
| 2 | Thurs. | 6:11 | 5:49 | 66 | 82 | 70 | mist fell til 9 a.m. |
| 3 | Fri. | 6:12 | 5:48 | 65 | 86 | 76 | exceedingly dusty |
| 4 | Sat. | 6:13 | 5:47 | 72 | 88 | 75 | slight shower at 6 p.m. |
| 5 | Sun. | 6:14 | 5:46 | 61 | 73 | 50 | |
| 6 | Mon. | 6:16 | 5:44 | 44 | 70 | 59 | |
| 7 | Tues. | 6:17 | 5:43 | 59 | 86 | 76 | a few high clouds |
| 8 | Wed. | 6:18 | 5:42 | 76 | 88 | 76 | |
| 9 | Thurs. | 6:19 | 5:41 | 64 | 86 | 66 | |
| 10 | Fri. | 6:21 | 5:39 | 63 | 74 | 74 | sprinkled at 1 p.m. |
| 11 | Sat. | 6:22 | 5:38 | 63 | 61 | 55 | sprinkly at 2 |
| 12 | Sun. | 6:23 | 5:37 | 50 | 56 | 54 | sprinkled all p.m. |
| 13 | Mon. | 6:24 | 5:36 | 50 | 63 | 58 | |
| 14 | Tues. | 6:26 | 5:34 | 55 | 71 | 62 | |
| 15 | Wed. | 6:27 | 5:33 | 55 | 69 | 62 | |
| 16 | Thurs. | 6:28 | 5:32 | 54 | 66 | 59 | |
| 17 | Fri. | 6:29 | 5:31 | 51 | 70 | 58 | heavy showers in night |
| 18 | Sat. | 6:30 | 5:30 | 44 | 62 | 48 | heavy dew |
| 19 | Sun. | 6:31 | 5:29 | 46 | 70 | 56 | |
| 20 | Mon. | 6:32 | 5:28 | 45 | 64 | 47 | |
| 21 | Tues. | 6:33 | 5:27 | 40 | 67 | 62 | heavy white frost |
| 22 | Wed. | 6:34 | 5:26 | 55 | 58 | 48 | no dew, wind all day |
| 23 | Thurs. | 6:36 | 5:24 | 42 | 54 | 42 | |
| 24 | Fri. | 6:37 | 5:23 | 34 | 60 | 52 | |
| 25 | Sat. | 6:38 | 5:22 | 45 | 72 | 53 | slight fog at 9 |
| 26 | Sun. | 6:39 | 5:21 | 48 | 52 | 54 | rained all day |
| 27 | Mon. | 6:40 | 5:20 | 44 | 50 | 43 | high wind all night |
| 28 | Tues. | 6:41 | 5:19 | 32 | 55 | 47 | |
| 29 | Wed. | 6:43 | 5:17 | 40 | 62 | 48 | |
| 30 | Thurs. | 6:44 | 5:16 | 34 | 62 | 52 | |
| 31 | Fri. | 6:45 | 5:15 | 44 | 71 | 58 | |

November 1862

On the night of November 6–7, "a heavy fall of snow" blanketed Richmond. The weather "continues cold" on the 9th, reading 38 degrees at an unspecified hour, and snow still on the ground. On the 20th, "It has been raining occasionally the last day or two," and rain continued all of that night.[82]

The "first snow of the season" near Millwood in the Shenandoah Valley on November 7 "commenced falling this morning and continued with but little intermission during the day." The morning of the 8th "cleared off beautifully, and the snow will soon disappear." A Virginia infantry captain outdoors in the storm on the 7th described its intensity: "It is snowing rapidly . . . the ground is already white, the wind howls and sweeps the chilly blasts around us, while the thick shower of snowflakes insure a shroud upon the wood and plains."[83]

The cold weather that invaded the Valley mid-month made maneuvers difficult. A soldier writing there on November 14 bewailed a recent march his "cold and hungry" comrades endured. As they crossed the Shenandoah River, "half waist deep, the water was freezing cold and the wind almost cutting you in two." The 14th, by happy contrast, was "a beautiful sunshiny" day.[84]

A Mississippi soldier hurrying toward Fredericksburg, where his brigade would stage a memorable defense of the riverfront, described a harrowing march: "It began to rain and sleet just before we reached Rapidan, where we forded the river and marched nearly all night in the cold and sleet. By next morning the weather had become bitter cold, and whenever we halted ten minutes our clothing would freeze on our bodies."[85]

"I think the Rapidan river was the coldest water I ever struck," a member of the 24th Georgia wrote of that same icy crossing. "After we got over we had to double-quick a half mile . . . barefoot and in our shirt-tails. . . . We had to march several miles in a cold rain."[86]

"A dull, heavy, sleeting rain" on November 19 soaked a Union engineer detachment heading from Washington toward Fredericksburg, "enough to make us all feel lonely, cheerless, desolate." The engineers continued their drenched march on the 20th through rain "falling in earnest," and again on a "cold, raw" 21st, "the rain . . . still pouring."[87]

Virginia horse artillerymen who arrived near Fredericksburg on the 20th spent a night of surpassing misery. "It had been raining all day and all hands were soaking wet and it was so cold the men could hardly take the harness off the horses." After walking around a sparse fire all night to keep from freezing, one shivering soldier declared: "Oh My Country! My Country! how can you repay me for this night's suffering." A Federal soldier across the river grumbled in his diary on the 21st that "very wet weather for a few days . . . makes our shelter tents very uncomfortable to live in."[88]

The threat of Federal bombardment of Fredericksburg sent thousands of civilians fleeing into the frigid countryside during late November. "It is indeed a sad & heartrending sight to see women & children driven from their homes," a South Carolinian wrote to his wife, "to take shelter in the woods with nothing but the broad canopy of Heaven for covering." Lieutenant Anderson J. Peeler of the 5th Florida Infantry described the pitiful refugees: "My memory can bring in view, no picture of such utter distress. . . . At night—in darkness, houseless, homeless . . . pierced through with cutting winds, were found weeping, wailing, distracted mothers with little children, whose tender little hands and feet were purple with biting cold, piteously crying."[89]

Federal soldiers sheltered in tents and with a supply of firewood could accept the early onset of winter more equably. A Pennsylvanian relished the "dry snow and keen, sharp air" on November 26, and considered the morning of the 27th "a grand and glorious one" because the ground, "lightly coated with snow made crisp by the previous evening's freezing," was no longer muddy.[90]

Civilians driven from Fredericksburg in November 1862 by the threat of shelling found scant refuge in the ice-encrusted countryside, "with nothing but the broad canopy of Heaven for covering." (Gettysburg National Military Park)

## November 1862

| Date | Day of Week | Rich-mond Sunrise | Rich-mond Sunset | D.C. Temperature | | | Other |
|---|---|---|---|---|---|---|---|
| | | | | 7 a.m. | 2 p.m. | 9 p.m. | |
| 1 | Sat. | 6:46 | 5:14 | 43 | 75 | 56 | halo around the moon |
| 2 | Sun. | 6:47 | 5:13 | 45 | 71 | 66 | |
| 3 | Mon. | 6:48 | 5:12 | 47 | 53 | 43 | wind all night |
| 4 | Tues. | 6:49 | 5:11 | 37 | 55 | 40 | |
| 5 | Wed. | 6:50 | 5:10 | 36 | 62 | 55 | drizzly at 9 p.m. |
| 6 | Thurs. | 6:51 | 5:09 | 44 | 45 | 37 | wind all night |
| 7 | Fri. | 6:52 | 5:08 | 33 | 38 | 30 | snowed 5 inches |
| 8 | Sat. | 6:53 | 5:07 | 29 | 39 | 34 | snow nearly gone |
| 9 | Sun. | 6:54 | 5:06 | 31 | 48 | 40 | |
| 10 | Mon. | 6:55 | 5:05 | 32 | 59 | 38 | |
| 11 | Tues. | 6:56 | 5:04 | 39 | 58 | 49 | white frost |
| 12 | Wed. | 6:57 | 5:03 | 50 | 63 | 56 | .16 last night |
| 13 | Thurs. | 6:58 | 5:02 | 47 | 59 | 42 | |
| 14 | Fri. | 6:59 | 5:01 | 33 | 60 | 46 | heavy white frost |
| 15 | Sat. | 7:00 | 5:00 | 41 | 56 | 42 | white frost |
| 16 | Sun. | 7:01 | 4:59 | 36 | 48 | 43 | snowlike |
| 17 | Mon. | 7:01 | 4:59 | 42 | 54 | 53 | slight shower |
| 18 | Tues. | 7:02 | 4:58 | 50 | 55 | 52 | foggy |
| 19 | Wed. | 7:03 | 4:57 | 51 | 58 | 62 | drizzly all day |
| 20 | Thurs. | 7:04 | 4:56 | 63 | 58 | 53 | thunder & lightning p.m. |
| 21 | Fri. | 7:05 | 4:55 | 44 | 47 | 42 | 1.78 last night |
| 22 | Sat. | 7:06 | 4:54 | 41 | 49 | 43 | |
| 23 | Sun. | 7:07 | 4:53 | 37 | 45 | 37 | |
| 24 | Mon. | 7:07 | 4:53 | 28 | 45 | 34 | |
| 25 | Tues. | 7:08 | 4:52 | 28 | 48 | 46 | heavy frost |
| 26 | Wed. | 7:09 | 4:51 | 40 | 43 | 37 | |
| 27 | Thurs. | 7:09 | 4:51 | 36 | 42 | 37 | |
| 28 | Fri. | 7:10 | 4:50 | 36 | 48 | 39 | |
| 29 | Sat. | 7:11 | 4:49 | 39 | 46 | 38 | |
| 30 | Sun. | 7:11 | 4:49 | 32 | 48 | 45 | |

## December 1862

Rain fell on Richmond the night of December 4 and into the 5th. A diarist in the city wrote on December 7: "Last night was bitter cold, and this morning there was ice on my wash-stand, within five feet of the fire. Is this the 'sunny South' the North is fighting to possess?" A week wrought a revolution, December 14 being "a bright and lovely Sabbath morning, and as warm as May." "A very cold day" in Richmond on December 20 had by 9 p.m. become "intensely frigid." Two days later: "the weather has greatly moderated."[91]

Soldiers near Fredericksburg, who soon would fight a great battle around that city, faced weather a bit worse than Richmond's. "Rain, Hail, Snow all day, and far into the night," a Delaware man wrote in his diary on December 5. Two days later he called the night of December 6–7 "the coldest I have experienced since I joined the Army. Three men were frozen to death on Picket." By the time of the battle, however, Indian summer had settled on central Virginia. "The weather has been splendid from the time we broke Camp on Thursday morning [December 11], except the mud," the Delaware soldier noted on the 15th. Apparently the pleasantly warm weather and general sunshine prompted him to ignore a brief shower early on the 14th. "A very heavy rain" on the night of December 15–16 broke the spell of lovely weather and served to cover the Federal retreat across the Rappahannock.[92]

The remarkable tendency of underinformed authors to declare that wounded men froze to death at Fredericksburg apparently began in the imagination of some fervid writer and has been aped mindlessly subsequently. The mild weather could not have frozen water during that period, much less human flesh. A local boy who rode the lines during the battle with his cavalry regiment reiterated what the weather readings prove. "The weather on the 13th of December, 1862, was much above freezing point," he recalled. "The roads were sloppy and muddy, but there was no snow on the ground except on the northern exposures and in the woods."[93]

"The morning of the 14th was ushered in by rain and dark clouds," a Virginian wrote of the weather at Fredericksburg the day after the great battle, "but early in the day the clouds lifted and the sun shone out."[94]

One of the most famous meteorological events of the war burst upon the horizon in full view of both armies near Fredericksburg on the evening of

December 14 in the form of "a glorious aurora borealis, like a red shield of honor." A New England soldier described the phenomenon in a letter home: "For over an hour the heavens were filled with long streaks of pale yellow light then all blended together and turned to a blood red & formed a complete fan shaped form . . . from NE to SW the sight was splendid."[95]

Artillerists posted near Port Royal during December 16–20 were "suffering very much from rain and cold winds."[96]

Late December in Fredericksburg brought "beautiful weather." A woman diarist near there called Christmas Day "almost springy."[97]

Christmas Eve in the camps outside Richmond was "bright and beautiful, but cold," a boy from Surry County wrote in a letter full of nostalgic yearnings for the joys of a holiday at home. He concluded his Christmas missive: "May the sun shine warm upon you at home."[98]

## December 1862

| Date | Day of Week | Rich-mond Sunrise | Rich-mond Sunset | D.C. Temperature | | | Other |
|---|---|---|---|---|---|---|---|
| | | | | 7 a.m. | 2 p.m. | 9 p.m. | |
| 1 | Mon. | 7:12 | 4:48 | 43 | 53 | 45 | |
| 2 | Tues. | 7:12 | 4:48 | 35 | 45 | 36 | murky, dim stars at 9 p.m. |
| 3 | Wed. | 7:13 | 4:47 | 36 | 41 | 34 | snowlike all day |
| 4 | Thurs. | 7:13 | 4:47 | 30 | 45 | 34 | at 2 p.m., snow & rain |
| 5 | Fri. | 7:14 | 4:46 | 35 | 36 | 33 | 2.00 snow last night |
| 6 | Sat. | 7:14 | 4:46 | 27 | 30 | 20 | |
| 7 | Sun. | 7:15 | 4:45 | 16 | 26 | 22 | |
| 8 | Mon. | 7:15 | 4:45 | 16 | 34 | 26 | |
| 9 | Tues. | 7:15 | 4:45 | 20 | 40 | 26 | |
| 10 | Wed. | 7:16 | 4:44 | 20 | 46 | 29 | |
| 11 | Thurs. | 7:16 | 4:44 | 24 | 50 | 32 ½ | lurid horizon evening |
| 12 | Fri. | 7:16 | 4:44 | 28 | 56 ½ | 31 | |
| 13 | Sat. | 7:17 | 4:43 | 34 | 56 | 40 | |
| 14 | Sun. | 7:17 | 4:43 | 40 | 63 | 50 ½ | |
| 15 | Mon. | 7:17 | 4:43 | 56 | 68 ½ | 60 | |
| 16 | Tues. | 7:17 | 4:43 | 40 | 43 | 36 | |
| 17 | Wed. | 7:18 | 4:42 | 30 | 33 | 30 | at 2 spits snow |
| 18 | Thurs. | 7:18 | 4:42 | 21 | 36 | 29 | |
| 19 | Fri. | 7:18 | 4:42 | 27 | 46 | 36 | |
| 20 | Sat. | 7:18 | 4:42 | 17 | 37 | 18 | |
| 21 | Sun. | 7:18 | 4:42 | 12 | 20 | 28 | Winter Solstice |
| 22 | Mon. | 7:18 | 4:42 | 28 | 39 | 36 | |
| 23 | Tues. | 7:18 | 4:42 | 39 | 59 | 41 | |
| 24 | Wed. | 7:18 | 4:42 | 38 | 45 | 44 | |
| 25 | Thurs. | 7:18 | 4:42 | 38 | 51 | 57 | |
| 26 | Fri. | 7:18 | 4:42 | 44 | 60 | 57 | raining at 7 p.m. |
| 27 | Sat. | 7:17 | 4:43 | 52 | 54 | 45 | |
| 28 | Sun. | 7:17 | 4:43 | 37 | 51 | 40 | |
| 29 | Mon. | 7:17 | 4:43 | 28 | 60 | 42 | heavy white frost |
| 30 | Tue | 7:17 | 4:43 | 37 | 48 | 45 | began to rain at 5 p.m. |
| 31 | Wed. | 7:17 | 4:43 | 40 | 39 | 36 | spit snow at 4 |

# Chapter 4

# 1863

## January 1863

The new year dawned in Richmond under gloomy skies, but they quickly gave way to sun that "beamed forth in great splendor." On January 16 it had been "blowing a gale for two days," and the "bitter cold" night of the 17th left "everything . . . frozen" in the Confederate capital city. Rain "fell in torrents" on the night of the 20th and continued the next day with "a violent storm of wind from the N.W." A big snowstorm hit Richmond on the night of the 27th, and "it snowed incessantly—melting rapidly, however," on the 28th.[1]

Artillerists in the Richmond defenses marched on a false alarm late on January 8, then went into camp under "cloudy and stormy-looking" skies. As they slept, "down thick and fast fell the fleecy snow, covering us all with a blanket of white." When the gunners awoke in a morning that "dawned bright and beautiful," the "novelty and romance" of the setting delighted the young soldiers.[2]

Union General Ambrose E. Burnside, hoping to retrieve his fortunes after a shattering defeat at Fredericksburg in December, began to move up the left bank of the Rappahannock River on January 19. Four days later he had been thoroughly routed by the elements in what came to be known derisively as the "Mud March." A Confederate diary meticulous about the weather reported the 19th as "very fine . . . bright sunshine and agreeably cool until late in the p.m. when a cold wind sprung up." More "chilly . . . biting wind" on the 20th brought in rain by 7 p.m., "quite hard, and . . . quite steadily"—"the wind blew like the 'old nick.'" On the 21st it rained "all day, and is still at it, very hard now, midnight . . . the water is standing over the country." A New York colonel disgustedly described the 21st as "about as stormy a day as it is possible to get up." On the 22nd "it rained all day" once again. One of the muddy, drenched Federal soldiers described Confederate reaction to the fiasco in his diary: "The Rebs across the river stuck up a board with the words 'Burnside stuck in the Mud' in large letters."[3] Burnside gave up, and soon afterward Lincoln cashiered him.

A Virginia soldier camped at Grace Church, south of Fredericksburg, grumbled in his diary on January 23: "Rain—rain—rain—nothing but rain from morning until night, and from night until morning. What a dull time." At the same camp, soldiers awoke on January 28 "to find a heavy snow had fallen during the night, and it continued throughout the day." The next morning "cleared off beautifully . . . the snow is melting fast."[4]

## January 1863

| Date | Day of Week | Rich-mond Sunrise | Rich-mond Sunset | D.C. Temperature | | | Other |
|------|-------------|-------------------|------------------|-------|--------|--------|-------|
| | | | | 7 a.m. | 2 p.m. | 9 p.m. | |
| 1 | Thurs. | 7:16 | 4:44 | 28 | 46 | 34 | a fine, clear day |
| 2 | Fri. | 7:16 | 4:44 | 24 | 48 | 35 | heavy white frost |
| 3 | Sat. | 7:15 | 4:45 | 24 | 49 | 34 | heavy white frost |
| 4 | Sun. | 7:15 | 4:45 | 40 | 58 | 50 | shower at 5 p.m. |
| 5 | Mon. | 7:15 | 4:45 | 35 | 56 | 42 | heavy white frost |
| 6 | Tues. | 7:14 | 4:46 | 34 | 44 | 39 | heavy white frost |
| 7 | Wed. | 7:14 | 4:46 | 37 | 31 | 22 | |
| 8 | Thurs. | 7:13 | 4:47 | 21 | 32 | 30 | snowlike all p.m. |
| 9 | Fri. | 7:12 | 4:48 | 30 | 40 | 35 | .50 snow last night |
| 10 | Sat. | 7:12 | 4:48 | 35 | 40 | 49 | .75 rain last night |
| 11 | Sun. | 7:11 | 4:49 | 32 | 50 | 40 | |
| 12 | Mon. | 7:11 | 4:49 | 29 | 45 | 37 | |
| 13 | Tues. | 7:10 | 4:50 | 33 | 48 | 39 | |
| 14 | Wed. | 7:09 | 4:51 | 39 | 47 | 58 | |
| 15 | Thurs. | 7:09 | 4:51 | 58 | 68 | 66 | windy last night |
| 16 | Fri. | 7:08 | 4:52 | 53 | 44 | 29 | .48 last night |
| 17 | Sat. | 7:07 | 4:53 | 21 | 30 | 22 | |
| 18 | Sun. | 7:06 | 4:54 | 13 | 22 | 24 | |
| 19 | Mon. | 7:06 | 4:54 | 22 | 38 | 33 | snowlike at 8 a.m. |
| 20 | Tues. | 7:05 | 4:55 | 29 | 36 | 36 | snowlike all day |
| 21 | Wed. | 7:04 | 4:56 | 36 | 42 | 38 | 1.25 rain last night & wind |
| 22 | Thurs. | 7:03 | 4:57 | 38 | 43 | 39 | rain all day .72 |
| 23 | Fri. | 7:03 | 4:57 | 38 | 50 | 44 | |
| 24 | Sat. | 7:02 | 4:58 | 40 | 52 | 46 | |
| 25 | Sun. | 7:01 | 4:59 | 43 | 54 | 41 | |
| 26 | Mon. | 7:00 | 5:00 | 42 | 59 | 54 | |
| 27 | Tues. | 6:59 | 5:01 | 54 | 47 | 42 | drizzled all p.m. |
| 28 | Wed. | 6:58 | 5:02 | 34 | 34 | 32 | |
| 29 | Thurs. | 6:57 | 5:03 | 31 | 39 | 32 | 6 inches snow last night |
| 30 | Fri. | 6:56 | 5:04 | 20 | 42 | 34 | snow, rain all day |
| 31 | Sat. | 6:55 | 5:05 | 32 | 52 | 37 | |

The campaign in the mud, January 1863. (Alfred H. Guernsey and Henry M. Alden, eds., *Harper's Pictorial History of the Civil War* [Chicago: Star Publishing Company, 1894], 418–19.)

*Continued on the next page*

The campaign in the mud, January 1863. *Continued*

## February 1863

A War Department bureaucrat in Richmond described "extremely cold weather" on the snowy night of February 4 and predicted that the rivers would freeze. He claimed that the thermometer on the night of the 3rd had dropped to 8 degrees below zero. February 14 in Richmond proved to be "a beautiful day." February 22, 1862, had been a miserably wet day for Jefferson Davis's inauguration, and the first anniversary proved even worse—"the ugliest day I ever saw," a diarist declared. "Snow fell all night, and was falling fast all day, with a northwest wind howling furiously. The snow is now nearly a foot deep, and the weather very cold." On the last day of the month, Richmond's weather "was too bad" for a renter to move his belongings into new quarters.[5]

Virginia horse artillerists marching toward Fredericksburg on February 16 complained of "rain, snow and the coldest kind of weather." On the 18th, "the ground was covered with snow and the mud was knee deep." After reaching Hamilton's Crossing, they awoke on the 22nd to find that "about eight inches of snow had fallen and covered us during the night" and that some horses had died of exposure.[6]

A soldier-diarist camped near Bowling Green described the weather there: February 11, "snow and rain"; 12th, "clear and cool"; 13th, 16th, "clear and pleasant"; 17th, "snowed all day. Cold"; 18th to 21st, "rainy and cold"; 22nd, "snow ten to twelve inches deep"; and rain on the 26th and 28th.[7]

The snow provided a frolic for an artillery staff officer, who enjoyed a visit in camp from some ladies on February 23: "Went down to the ice pond with the ladies, snowballed. . . . Built snow man, statue of Gen'l. Washington."[8]

General Fitzhugh Lee's Confederates crossed Kelly's Ford on the night of February 24 and won a skirmish at Hartwood Church the next day. A trooper who accompanied the column recalled snow "five inches deep" during the night ride.[9]

Snowball battles in Confederate camps sometimes expanded to large-scale encounters between units. Many soldiers from the Deep South had never seen snow before coming to Virginia with their regiments. (Robert Underwood Johnson and C. C. Buel, eds., *Battles and Leaders of the Civil War: Being for the Most Part Contributions by Union and Confederate Officers,* 4 vols. [New York: The Century Co., 1887–88], 3:99.)

February 1863

| Date | Day of Week | Rich- mond Sunrise | Rich- mond Sunset | D.C. Temperature | | | Other |
|------|------|------|------|------|------|------|------|
| | | | | 7 a.m. | 2 p.m. | 9 p.m. | |
| 1 | Sun. | 6:54 | 5:06 | 31 | 52 | 48 | began to drizzle 5 p.m. |
| 2 | Mon. | 6:53 | 5:07 | 36 | 38 | 28 | |
| 3 | Tues. | 6:52 | 5:08 | 24 | 27 | 14 | |
| 4 | Wed. | 6:51 | 5:09 | 10 | 24 | 18 | |
| 5 | Thurs. | 6:50 | 5:10 | 16 | 28 | 42 | |
| 6 | Fri. | 6:49 | 5:11 | 42 | 34 | 30 | spits snow all day |
| 7 | Sat. | 6:48 | 5:12 | 28 | 40 | 32 | last night .75 |
| 8 | Sun. | 6:47 | 5:13 | 28 | 50 | 40 | snowlike at 9 p.m. |
| 9 | Mon. | 6:46 | 5:14 | 36 | 57 | 43 | white frost |
| 10 | Tues. | 6:45 | 5:15 | 41 | 60 | 40 | |
| 11 | Wed. | 6:44 | 5:16 | 35 | 40 | 35 | sprinkling at 2 p.m. |
| 12 | Thurs. | 6:43 | 5:17 | 36 | 56 | 42 | |
| 13 | Fri. | 6:42 | 5:18 | 34 | 44 | 32 | |
| 14 | Sat. | 6:41 | 5:19 | 39 | 43 | 37 | |
| 15 | Sun. | 6:40 | 5:20 | 42 | 56 | 44 | drizzly 7 to noon |
| 16 | Mon. | 6:39 | 5:21 | 33 | 51 | 39 | at 10 snowlike |
| 17 | Tues. | 6:37 | 5:23 | 34 | 37 | 33 | snow last night & all day til 4 p.m. (8 inches) |
| 18 | Wed. | 6:36 | 5:24 | 32 | 37 | 37 | |
| 19 | Thurs. | 6:35 | 5:25 | 37 | 44 | 43 | drizzle all day |
| 20 | Fri. | 6:34 | 5:26 | 45 | 52 | 40 | |
| 21 | Sat. | 6:32 | 5:28 | 38 | 42 | 33 | |
| 22 | Sun. | 6:31 | 5:29 | 20 | 26 | 22 | snow all day 9 inches |
| 23 | Mon. | 6:30 | 5:30 | 19 | 32 | 29 | |
| 24 | Tues. | 6:29 | 5:31 | 22 | 41 | 28 | |
| 25 | Wed. | 6:28 | 5:32 | 20 | 45 | 34 | halo about the moon |
| 26 | Thurs. | 6:27 | 5:33 | 39 | 42 | 42 | |
| 27 | Fri. | 6:26 | 5:34 | 47 | 63 | 47 | |
| 28 | Sat. | 6:25 | 5:35 | 41 | 50 | 40 | |

## March 1863

Richmond enjoyed "fine March weather" on the 3rd, "but the floods of late have damaged the railroad bridges between this and Fredericksburg." A "violent snow-storm" blanketed Richmond on the 10th. A "cold, disagreeable day" on the 15th and sleet on the 16th prompted a diarist to declare that "March so far has been as cold and terrible as a winter month." More snow fell on the capital on the 19th, and the next morning it was "eight inches deep . . . and it is still falling fast." The snow accumulated to a foot by the time it tapered off on the 21st. A warm front produced fog on the 22nd, and by the next day the thaw had quickly turned the layer of snow into a vast muddy quagmire.[10]

Sunday, March 8, was "a bright and beautiful day" in the army camps surrounding Richmond.[11]

The 17th Virginia Infantry of General George E. Pickett's division, located in the vicinity of Hanover Junction on March 17, faced "a heavy snowstorm" that turned into rain about 10 p.m. "The rain continued through the entire night and the following day."[12]

Early March near Fredericksburg brought "a little rain and wind, then a li tle more rain and wind, then much more of both, until the 10th." Artillerists camped near Kelly's Ford from March 17 until April 13 "had a good deal of snow and rain. . . . March reversed its usual order this year. . . . It came in like a lion and went out like one, but not a dead one by any means."[13]

On March 17, as Major John Pelham suffered his mortal wound at Kelly's Ford, the Georgetown station reported no clouds and light southwest winds. Recent weather affected the field where Kelly's Ford was fought. "There had been a heavy rain just before this which finished up with a snow about five inches deep," a member of the Second Virginia Cavalry reported. "This was still on the ground, but the river was swollen from the rain until it was deep fording on horse back. . . . When we reached the breast works at the river, some of the rifle pits were filled with snow and ice."[14]

The *Richmond Sentinel* reported the big mid-month storm in emphatic terms: "The 19th and 20th of March will be recorded as the days on which the greatest snow storm of 1863 visited the region of Richmond. From half past 4 o'clock Thursday afternoon [19th] up to yesterday [20th] at 3, it fell almost incessantly, and covered the ground to a depth of eight or ten inches." In the midst of that blizzard, Major John Pelham's remains arrived at the capitol to be laid in state.[15]

A Virginia infantryman camped near Petersburg reported on the pervasive bad weather beginning on March 19: "Another heavy snowstorm set in on the evening of the 19th and the snow kept falling all night, all day and night of the 20th, and until the morning of the 21st, when it turned into a drizzling rain. There were fourteen or fifteen inches of snow on the ground." A general coping with the weather on the 21st called it "very disagreeable"—a word often used to describe the unpleasant weather of this month.[16]

When Congressman Franklin B. Sexton of Texas awoke in Richmond on March 20, he found "snow several inches deep . . . & falling heavily. Fell nearly all day & at night is a foot deep or more."[17] The same storm dumped "a very deep snow" that covered Botetourt County on March 21.[18]

In a Caroline County camp on March 28, "rain, rain, rain,—so long as this weather continues there is no chance for active operations."[19]

March 1863

| Date | Day of Week | Rich-mond Sunrise | Rich-mond Sunset | D.C. Temperature | | | Other |
|------|-------------|-------------------|------------------|-------|-------|-------|-------|
| | | | | 7 a.m. | 2 p.m. | 9 p.m. | |
| 1 | Sun. | 6:23 | 5:37 | 40 | 42 | 45 | |
| 2 | Mon. | 6:22 | 5:38 | 37 | 54 | 42 | |
| 3 | Tues. | 6:21 | 5:39 | 38 | 42 | 35 | 3 p.m. snowgust & thunder |
| 4 | Wed. | 6:20 | 5:40 | 38 | 33 | 27 | |
| 5 | Thurs. | 6:19 | 5:41 | 19 | 35 | 31 | |
| 6 | Fri. | 6:17 | 5:43 | 34 | 57 | 44 | sleety hail at daybreak |
| 7 | Sat. | 6:16 | 5:44 | 43 | 55 | 44 | rain |
| 8 | Sun. | 6:15 | 5:45 | 38 | 50 | 46 | .57 yesterday & night |
| 9 | Mon. | 6:14 | 5:46 | 38 | 54 | 43 | snows slightly all day |
| 10 | Tues. | 6:12 | 5:48 | 35 | 36 | 36 | |
| 11 | Wed. | 6:11 | 5:49 | 36 | 48 | 38 | snow 1.5 last night |
| 12 | Thurs. | 6:10 | 5:50 | 31 | 38 | 28 | snow shower at 2 |
| 13 | Fri. | 6:09 | 5:51 | 25 | 30 | 27 | |
| 14 | Sat. | 6:08 | 5:52 | 25 | 41 | 35 | |
| 15 | Sun. | 6:07 | 5:53 | 35 | 34 | 28 | thunder at 4 p.m. |
| 16 | Mon. | 6:06 | 5:54 | 26 | 35 | 32 | |
| 17 | Tues. | 6:04 | 5:56 | 28 | 48 | 42 | spits snow, smoky |
| 18 | Wed. | 6:03 | 5:57 | 42 | 46 | 37 | |
| 19 | Thurs. | 6:02 | 5:58 | 31 | 37 | 34 | |
| 20 | Fri. | 6:00 | 6:00 | 31 | 38 | 33 | Vernal Equinox |
| 21 | Sat. | 5:59 | 6:01 | 28 | 40 | 36 | snow 1.5 last night |
| 22 | Sun. | 5:58 | 6:02 | 38 | 56 | 52 | |
| 23 | Mon. | 5:57 | 6:03 | 44 | 56 | 46 | |
| 24 | Tues. | 5:55 | 6:05 | 43 | 59 | 52 | |
| 25 | Wed. | 5:54 | 6:06 | 57 | 70 | 48 | .81 last night |
| 26 | Thurs. | 5:53 | 6:07 | 41 | 44 | 39 | |
| 27 | Fri. | 5:52 | 6:08 | 36 | 53 | 44 | halo around moon |
| 28 | Sat. | 5:51 | 6:09 | 41 | 44 | 40 | .81 rained all day, thunder at 10:30 |
| 29 | Sun. | 5:50 | 6:10 | 34 | 38 | 32 | |
| 30 | Mon. | 5:49 | 6:11 | 32 | 50 | 44 | |
| 31 | Tues. | 5:48 | 6:12 | 34 | 35 | 40 | snowed all night, 2 inches |

April 1863

In Virginia's Tidewater, April came in unusually cool. A woman's diary from Norfolk County reported "fair and a cold north wind" on the 1st; "cold and cloudy" on the 2nd, "still cold and cloudy" on the 3rd, and "cold and cloudy and blustering" on the 4th. Not until the 11th did she report "a lovely spring day."[20]

"To-day has been cold and stormy," a Fairfax County woman wrote in her diary on April 2. "It rained all the early part of it, and the wind has blown a perfect hurricane." She reported skies "cold and cloudy . . . dull and dreary" on the 7th, and on the 12th she complained: "It has been so long since we had any pleasant weather," but this day turned "bright, and warm."[21]

Spring put in an appearance near Front Royal as the month began. April 1 was "bright and pleasant but quite cold," and April 2 "clear and warm but quite windy." Then April 3 dawned as "a lovely morning such as we've not had in a long time."[22]

Snow in Richmond on the night of April 4–5 amounted to "several inches" by morning, but rain soon melted the accumulation. The second anniversary of Fort Sumter brought "warm weather at last, and dry." "A seasonal rain" fell on the capital city on the 15th. By the 18th, in consequence of the "lovely weather . . . vegetation shows signs of the return of the vernal season."[23]

A Mississippi soldier encamped near Fredericksburg awoke on April 5 to find snow "six or seven inches deep this morning," but a warm sun melted it "fast." The snowfall was "the fourth one since I got back" from a furlough in early March, he reported.[24]

At daybreak on April 15 "it was raining so hard" along the northern Virginia military frontier that a Federal cavalry movement was postponed, "and at 12 o'clock it was raining harder than ever." Eventually the column started off "in the hardest kind of rain."[25] The same storm pelted Confederate camps near Fredericksburg, prompting an Alabama captain to bewail that kind of weather: "This is one of the dreaded days in camp, raining heavy all day, all closely confined to their several or respective bunk or shelters."[26]

In Madison County, rain began on the night of April 19 and the next morning "the clouds are very heavy, enveloping the mountains completely." Later on the 20th the rain redoubled into "a driving storm."[27]

On April 23, a woman living in Botetourt County wrote of "enjoying . . . the beautiful spring weather" of recent days.[28]

April 1863

| Date | Day of Week | Rich-mond Sunrise | Rich-mond Sunset | D.C. Temperature | | | Other |
|------|------|------|------|------|------|------|------|
| | | | | 7 a.m. | 2 p.m. | 9 p.m. | |
| 1 | Wed. | 5:46 | 6:14 | 31 | 38 | 37 | |
| 2 | Thurs. | 5:45 | 6:15 | 40 | 68 | 58 | sprinkled at 6 p.m. |
| 3 | Fri. | 5:44 | 6:16 | 41 | 58 | 45 | white frost last night |
| 4 | Sat. | 5:43 | 6:17 | 30 | 42 | 32 | |
| 5 | Sun. | 5:41 | 6:19 | 32 | 41 | 39 | Easter. 12-inch snow last night, heavy til noon |
| 6 | Mon. | 5:40 | 6:20 | 38 | 50 | 42 | |
| 7 | Tues. | 5:39 | 6:21 | 37 | 56 | 40 | |
| 8 | Wed. | 5:38 | 6:22 | 36 | 46 | 40 | |
| 9 | Thurs. | 5:36 | 6:24 | 38 | 52 | 44 | |
| 10 | Fri. | 5:35 | 6:25 | 35 | 66 | 52 | white frost |
| 11 | Sat. | 5:34 | 6:26 | 48 | 72 | 40 | |
| 12 | Sun. | 5:33 | 6:27 | 57 | 71 | 59 | |
| 13 | Mon. | 5:32 | 6:28 | 48 | 54 | 45 | 1.05 shower last night |
| 14 | Tues. | 5:31 | 6:29 | 36 | 62 | 51 | white frost |
| 15 | Wed. | 5:30 | 6:30 | 43 | 48 | 50 | rained all night |
| 16 | Thurs. | 5:29 | 6:31 | 48 | 52 | 49 | 1.17 yesterday & night, rained all day |
| 17 | Fri. | 5:28 | 6:32 | 50 | 64 | 54 | .16 last night |
| 18 | Sat. | 5:26 | 6:34 | 52 | 70 | 55 | |
| 19 | Sun. | 5:25 | 6:35 | 59 | 70 | 51 | |
| 20 | Mon. | 5:24 | 6:36 | 53 | 56 | 50 | rained all p.m., .22 |
| 21 | Tues. | 5:23 | 6:37 | 43 | 52 | 49 | drizzly 7 p.m. |
| 22 | Wed. | 5:22 | 6:38 | 44 | 64 | 54 | |
| 23 | Thurs. | 5:21 | 6:39 | 46 | 52 | 51 | rained all day |
| 24 | Fri. | 5:20 | 6:40 | 48 | 54 | 54 | .84 last night & today |
| 25 | Sat. | 5:19 | 6:41 | 49 | 60 | 50 | |
| 26 | Sun. | 5:18 | 6:42 | 44 | 60 | 42 | peaches in full bloom |
| 27 | Mon. | 5:17 | 6:43 | 48 | 72 | 57 | plum trees in bloom |
| 28 | Tues. | 5:16 | 6:44 | 52 | 62 | 59 | .33, rained all p.m. |
| 29 | Wed. | 5:15 | 6:45 | 58 | 74 | 60 | sprinkled |
| 30 | Thurs. | 5:14 | 6:46 | 55 | 64 | 57 | cherry trees in bloom, shower at dusk |

## May 1863

General Lee's Army of Northern Virginia went into battle at Chancellorsville on May 1 under "quite pleasant" weather conditions, "a genuine May day." On May 2, while "Stonewall" Jackson and nearly thirty thousand men executed that general's most famous (and final) march of his career, "the day was quite warm and pleasant—the night clear. The trees are becoming quite green and the apple and pear trees are in full bloom."[29]

A full moon lit northern Virginia on the evening of May 2 as Jackson rode in front of his lines near Chancellorsville and met with a fatal volley fired by maladroit North Carolinians.[30]

Clear skies and bright sun on May 3 helped to place the time of one of the most famous photographs of Confederate dead taken during the war, after a rearguard action in Fredericksburg. Photographer A. J. Russell took pictures of dead men from the 18th Mississippi in the Sunken Road, immediately after Federals captured the area at about noon. The noon sun, squarely overhead, casts shadows directly below rifles lying above a small trench beside the road, confirming the timing as immediately after the late-morning attack. The Georgetown weather ledger confirms the absence of cloud cover on May 3.

On the afternoon of May 5, as Union General Joseph Hooker contemplated abandoning his bridgehead near Chancellorsville, "rain fell in torrents and soon our trenches were filled with water," a Confederate gunner wrote in his diary. "The cold rain perfectly benumbed us. . . . Little streams in a few moments [became] roaring torrents."[31]

Federal wounded lying helpless beneath the downpour on May 5 suffered under "terrific" thunder and lightning, "and the rain came down in sheets." A New Yorker declared that he had "never before witnessed and certainly never experienced one like it." Wounded men in the Chancellorsville clearing "lay sprawled in the mud and filth . . . in from three to five inches of water." Two of them drowned.[32]

May 12, the day of Jackson's funeral in Richmond, duplicated the 11th's weather: "bright and excessively hot." Several days leading up to May 17 were "cool and dry." A week later, a diarist in the capital boasted of "a fortnight of

Confederate dead in the Sunken Road, Fredericksburg, near noon on May 3, 1863. The bright midday sun on a cloudless day makes possible confirmation of the time of the photograph, since the shadows appear directly beneath the rifles. (Fredericksburg and Spotsylvania National Military Park)

calm, dry, and warm weather. There is a hazy atmosphere, and the sun rises and sets wearing a blood-red aspect. . . . It is like Indian summer in May. The ground is dry and crusted."[33]

Additional columns in the Georgetown weather ledger provide details about clouds and wind that usually have not been reproduced in the tables in this book. For the three primary days of the Battle of Chancellorsville, those columns showed very few clouds and no measurable wind on May 1; some clouds and light wind from the south on May 2; and few clouds, wind from the south, on May 3. As Jackson lay dying at Guiney Station on May 10, Georgetown reported a negligible wind from the south and a few light clouds.

May 12 in Lynchburg turned "very hot; thermometer at eighty-eight degrees." A member of the committee that accompanied Jackson's corpse from

A noisy storm helped to cover Union General Joseph Hooker's retreat across the Rappahannock River at United States Ford on the night of May 5–6, 1863. (Robert Underwood Johnson and C. C. Buel, eds., *Battles and Leaders of the Civil War: Being for the Most Part Contributions by Union and Confederate Officers,* 4 vols. [New York: The Century Co., 1887–88], 3:222.)

Lynchburg to Lexington reported that "the morning threatened rain" on May 14, "and there was indeed several showers, but it cleared off." On May 23, back in Lynchburg, "weather warm. For several days we had a peculiar smoky atmosphere. The sun rises and sets 'shorn of its horizontal beams,' red as blood."[34]

Virginians voted in statewide elections on the 27th (electing General William "Extra Billy" Smith governor). "The day was quite warm" near Fredericksburg for the election, following seven consecutive days reported as "very warm."[35]

The month ended with strong winds on May 30–31. "The wind commenced to blow yesterday," a Fairfax woman wrote on the 31st, "and increased until to day it is a furious hurricane. The earth looks dry and parched, and the young green foliage is thrashed, and torn."[36]

May 1863

| Date | Day of Week | Richmond Sunrise | Richmond Sunset | D.C. Temperature | | | Other |
|------|-------------|-------------------|------------------|------|-------|-------|-------|
| | | | | 7 a.m. | 2 p.m. | 9 p.m. | |
| 1 | Fri. | 5:13 | 6:47 | 52 | 74 | 63 | heavy fog & dew |
| 2 | Sat. | 5:12 | 6:48 | 57 | — | 63 | |
| 3 | Sun. | 5:11 | 6:49 | 55 | 80 | 65 | |
| 4 | Mon. | 5:10 | 6:50 | 65 | 76 | 65 | showery, thunder & lightning at 8:30 |
| 5 | Tues. | 5:09 | 6:51 | 60 | 76 | 52 | .56 last night. Began to rain at 5 p.m. |
| 6 | Wed. | 5:08 | 6:52 | 44 | 50 | 45 | 1.43 yesterday p.m. & night, rained til 2 p.m. |
| 7 | Thurs. | 5:07 | 6:53 | 43 | 50 | 49 | .75 last night, drizzly all day |
| 8 | Fri. | 5:06 | 6:54 | 48 | 50 | 58 | drizzled til 2:30 p.m. |
| 9 | Sat. | 5:05 | 6:55 | 52 | 72 | 62 | |
| 10 | Sun. | 5:04 | 6:56 | 59 | 85 | 70 | |
| 11 | Mon. | 5:03 | 6:57 | 52 | 88 | 77 | |
| 12 | Tues. | 5:02 | 6:58 | 66 | — | 79 | |
| 13 | Wed. | 5:01 | 6:59 | 71 | 86 | 64 | |
| 14 | Thurs. | 5:00 | 7:00 | 62 | 70 | 60 | .34 last night |
| 15 | Fri. | 4:59 | 7:01 | 55 | 70 | 58 | |
| 16 | Sat. | 4:58 | 7:02 | 56 | 65 | 65 | dim Sun |
| 17 | Sun. | 4:57 | 7:03 | 65 | 79 | 64 | |
| 18 | Mon. | 4:57 | 7:03 | 50 | 70 | 57 | |
| 19 | Tues. | 4:56 | 7:04 | 60 | 86 | 64 | |
| 20 | Wed. | 4:55 | 7:05 | 58 | 84 | 70 | |
| 21 | Thurs. | 4:55 | 7:05 | 66 | — | 69 | |
| 22 | Fri. | 4:54 | 7:06 | 73 | 90 | 71 | smoky & extremely dusty |
| 23 | Sat. | 4:53 | 7:07 | 72 | 92 | 72 | |
| 24 | Sun. | 4:53 | 7:07 | 70 | — | 82 | |
| 25 | Mon. | 4:52 | 7:08 | 55 | 66 | 59 | drizzly & heavy mist |
| 26 | Tues. | 4:51 | 7:09 | 55 | 68 | 62 | |
| 27 | Wed. | 4:51 | 7:09 | 58 | 76 | 65 | |
| 28 | Thurs. | 4:50 | 7:10 | 58 | 78 | 66 | |
| 29 | Fri. | 4:49 | 7:11 | 62 | 82 | 73 | clear sky |
| 30 | Sat. | 4:49 | 7:11 | 75 | 80 | 76 | |
| 31 | Sun. | 4:48 | 7:12 | 74 | 79 | 72 | fine shower at 1 p.m. |

June 1863

A long dry spell in Richmond broke on June 4: "After a month of dry weather, we have just had a fine rain, most refreshing." The capital city basked under "a beautiful, bright warm summer day" on the 12th. Vegetable gardens in and around the city languished "in consequence of the protracted dry weather," but a diary entry on the 22nd applauded "at last, abundant rains." Six days later the diarist reported "copious rains recently" and "a cool, cloudy day."[37]

In Fairfax County a woman diarist wrote on June 8: "The weather for the past three or four weeks has been extremely dry and windy, every thing parched up. The grass as brown and crisp as it usually is in August."[38]

As the intense cavalry clash at Brandy Station unfolded on June 9, Georgetown reported no clouds at all, and wind from the northwest and west. An Alabama soldier who marched near Culpeper on the 9th called the day "very warm" and remarked on the number of men who fell out of ranks on the march because of the dire effects of the heat.[39]

The noted English visitor Arthur J. L. Fremantle arrived at Richmond on June 18 and found the weather "extremely hot and oppressive . . . for the first time since I left Havana, I really suffered from the heat." As he approached Culpeper on the 20th, he wrote: "quite cool after the rain of last night." Near Winchester on the 25th, Fremantle encountered "cool and showery" conditions. Before dawn on the 26th he awoke in Martinsburg to "drenching rain," then skies became "a little clearer" at 2 p.m.[40]

General J. E. B. Stuart and his cavalry fought intense combats against mounted Federals in mid-June along the eastern foreslopes of the Blue Ridge. A woman living nearby described June 16 as "fine and bracing." On the 20th, she sympathized with Confederates coping with "an inglorious rain." That night she called "one of blackness."[41]

Some Federals marching toward a date with destiny at Gettysburg welcomed the rain of the night of June 19–20 as a relief from the heat. A Maine soldier called it "a most delicious rainstorm. . . . We were wet to the skin, but slept nicely."[42]

The journal of British officer A. J. L. Fremantle reported his impressions of the weather in June and July 1863 at Richmond and during the Gettysburg campaign. (Dr. Mike Masters)

Confederates who entered Frederick, Maryland, on Saturday the 20th encountered "wet and rainy weather," but "Sunday dawned bright and clear."[43]

A Virginia cavalryman riding on Stuart's ill-starred raid en route to Gettysburg described the night of June 25, in a contemporary letter, as "rainy and disagreeable."[44]

Temperature readings at Gettysburg, not surprisingly, usually registered slightly lower than in Georgetown at the end of June (reading times, as usual, were 7 a.m., 2 p.m., and 9 p.m.): 59, 51, and 63 on the 25th; 60, 63, and 62 on the 26th; 61, 63, and 67 on the 27th; 63, 67, and 68 on the 28th; 66, 72, and 69 on the 29th; and 68, 79, and 71 on the 30th.[45]

## June 1863

| Date | Day of Week | Rich-mond Sunrise | Rich-mond Sunset | D.C. Temperature | | | Other |
|------|-------------|-------------------|------------------|------------------|------|------|-------|
| | | | | 7 a.m. | 2 p.m. | 9 p.m. | |
| 1 | Mon. | 4:48 | 7:12 | 74 | 86 | 73 | |
| 2 | Tues. | 4:47 | 7:13 | 70 | 86 | 66 | |
| 3 | Wed. | 4:47 | 7:13 | 60 | 76 | 63 | slight shower at night |
| 4 | Thurs. | 4:46 | 7:14 | 59 | 86 | 66 | |
| 5 | Fri. | 4:46 | 7:14 | 66 | — | 69(?) | sprinkly at 7 p.m. |
| 6 | Sat. | 4:45 | 7:15 | 66 | 86 | 65 | few drops at 6:30 |
| 7 | Sun. | 4:45 | 7:15 | 63 | 70 | 57 | |
| 8 | Mon. | 4:44 | 7:16 | 5- | 66 | 58 | |
| 9 | Tues. | 4:44 | 7:16 | 57 | 82 | 71 | |
| 10 | Wed. | 4:44 | 7:16 | 64 | 88 | 67 | |
| 11 | Thurs. | 4:44 | 7:16 | 72 | 80 | 74 | sprinkled at 2 p.m. |
| 12 | Fri. | 4:43 | 7:17 | 73 | 84(?) | 66 | |
| 13 | Sat. | 4:43 | 7:17 | 66 | 86 | 64 | .73, much thunder & lightning |
| 14 | Sun. | 4:43 | 7:17 | 66 | 76 | 70 | |
| 15 | Mon. | 4:43 | 7:17 | 64 | 92 | 80 | |
| 16 | Tues. | 4:42 | 7:18 | 72 | 94 | 73 | |
| 17 | Wed. | 4:42 | 7:18 | 70 | 94 | 85 | |
| 18 | Thurs. | 4:42 | 7:18 | 82 | 96 | 70 | .81, hail, thunder, wind |
| 19 | Fri. | 4:42 | 7:18 | 70 | 81 | 68 | |
| 20 | Sat. | 4:42 | 7:18 | 60 | 70 | 64 | .14 last night |
| 21 | Sun. | 4:42 | 7:18 | 65 | 61 | 70(?) | Summer Solstice |
| 22 | Mon. | 4:42 | 7:18 | 65 | 76 | 68 | |
| 23 | Tues. | 4:42 | 7:18 | 65 | 79 | 70 | |
| 24 | Wed. | 4:42 | 7:18 | 64 | 74 | 70 | |
| 25 | Thurs. | 4:42 | 7:18 | 65 | 72 | 66 | drizzly at 7 |
| 26 | Fri. | 4:42 | 7:18 | 65 | 76 | 66 | rained all day to 7:30 p.m. |
| 27 | Sat. | 4:43 | 7:17 | 64 | — | 72 | |
| 28 | Sun. | 4:43 | 7:17 | 68 | 76 | 64 | |
| 29 | Mon. | 4:43 | 7:17 | 70 | 80 | 73 | |
| 30 | Tues. | 4:43 | 7:17 | 70 | 83 | 76 | |

## July 1863

As the armies fought a climactic battle at Gettysburg at the beginning of the month, Georgetown reported cloudy skies every day during July 1–4, with winds from the south on the first three days, and switching during July 4 from southeast, to south, to northeast.[46]

Readings taken at Gettysburg during the battle, recorded at the customary three hours, were 72, 76, 74 on July 1; 74, 81, 76 on July 2; and 73, 87, 76 on July 3. The 2 p.m. reading on July 3, as Pickett's Charge made its desperate way forward, proved to be the highest at Gettysburg during the entire month of July. A heavy storm on July 4 knocked temperature readings down to 69, 72, and 70 degrees that day.[47]

The downpours that so encumbered Confederate supply trains and ambulances during the retreat from Gettysburg "began to descend in torrents" at 1 p.m. on July 4, according to the diary of A. J. L. Fremantle. A fresh downpour began at 9 p.m., and that night "was very bad—thunder and lightning, torrents of rain—the road knee-deep in mud and water."[48]

In a letter dated July 13, a Virginia cavalryman described rain "descending in torrents, so dampening my paper as to render it almost useless," even though he was writing under the protection of a tent.[49]

Local defense troops mustered in the Richmond trenches at the beginning of July spent five days of watchful waiting in "fine weather" that left them "sun-burnt and covered with dust." July 6 broke the dry spell by "raining furiously." After the storm, "for several days" the sky continued "without a cloud . . . the red sun, dimly seen through the mist (at noonday), casts a baleful light on the earth." On July 14, at 5:30 p.m., "a heavy thunder-storm, accompanied with a deluging rain," struck the capital. After the dangerously dry spring of 1863, Richmond's weather remained steadily wet through the last part of July: "weather is bad . . . it has been raining nearly a month. . . . Still raining! The great fear is that the crops will be ruined."[50]

An Alexandria weather station reported 75 degrees at 2 p.m. on July 13, and 85 at 2 p.m. on the 21st, filling in blanks in the Georgetown ledger. It also indicated rain falling on July 3 from 3 a.m. to 4 p.m. For the month, the

Torrential downpours, and the mud they generated, added to the ordeal of the thousands of Confederate wounded retreating from Gettysburg in early July 1863. (Robert Underwood Johnson and C. C. Buel, eds., *Battles and Leaders of the Civil War: Being for the Most Part Contributions by Union and Confederate Officers,* 4 vols. [New York: The Century Co., 1887–88], 3:426.)

Alexandria recorder tabulated 16.67 days of fair weather, 14.33 days of cloudy weather, and 4.33 days of rain.[51]

Late on the night of the 16th, Confederate horse artillerists bivouacked near Leetown "in a heavy rain."[52]

In the lower Shenandoah Valley, where the Army of Northern Virginia recuperated from its campaign in Pennsylvania, temperatures were "very warm" on July 18 and remained hot with no mention of rain for ten days. On July 23, blackberries in the Valley were "getting ripe and are quite nice now." The next day, "our men suffered much" because of excessive heat. Finally, "a heavy shower" fell on Lee's parched army during the afternoon of July 30.[53]

## July 1863

| Date | Day of Week | Rich-mond Sunrise | Rich-mond Sunset | D.C. Temperature | | | Other |
|------|------|------|------|------|------|------|------|
| | | | | 7 a.m. | 2 p.m. | 9 p.m. | |
| 1 | Wed. | 4:43 | 7:17 | 70 | 89 | 78 | |
| 2 | Thurs. | 4:44 | 7:16 | 73 | 87 | 77 | |
| 3 | Fri. | 4:44 | 7:16 | 76 | 88 | 80 | shower at 4:30 p.m. |
| 4 | Sat. | 4:44 | 7:16 | 74 | 87 | 74 | .68 last night |
| 5 | Sun. | 4:45 | 7:15 | 72 | 86 | 76 | sprinkled some |
| 6 | Mon. | 4:45 | 7:15 | 78 | 84 | 70 | drizzly |
| 7 | Tues. | 4:45 | 7:15 | 72 | 80 | 77 | |
| 8 | Wed. | 4:46 | 7:14 | 72 | 80 | 73 | 1.36 yesterday & today, cleared at 3 p.m. |
| 9 | Thurs. | 4:46 | 7:14 | 74 | 82 | 73 | |
| 10 | Fri. | 4:47 | 7:13 | 73 | 94 | 76 | lightning in W at 9 p.m. |
| 11 | Sat. | 4:47 | 7:13 | 74 | 90 | 80 | |
| 12 | Sun. | 4:48 | 7:12 | 77 | 83 | 75 | |
| 13 | Mon. | 4:48 | 7:12 | 69 | -0 | 73 | last night & today 1.36 |
| 14 | Tues. | 4:49 | 7:11 | 76 | 84 | 77 | raining |
| 15 | Wed. | 4:50 | 7:10 | 75 | 84 | 76 | 1.52 last night |
| 16 | Thurs. | 4:50 | 7:10 | 75 | 82 | 70 | lightning, thunder at 7 p.m. |
| 17 | Fri. | 4:51 | 7:09 | 66 | 70 | 66 | .76 last night, showery all day |
| 18 | Sat. | 4:52 | 7:08 | 66 | 78 | 70 | .14 last night |
| 19 | Sun. | 4:52 | 7:08 | 69 | 81 | 74 | |
| 20 | Mon. | 4:53 | 7:07 | 72 | 87 | 80 | |
| 21 | Tues. | 4:54 | 7:06 | 75 | — | 73 | |
| 22 | Wed. | 4:55 | 7:05 | 67 | 80 | 69 | |
| 23 | Thurs. | 4:55 | 7:05 | 7- | 86 | 78 | |
| 24 | Fri. | 4:56 | 7:04 | 74 | — | 78 | |
| 25 | Sat. | 4:57 | 7:03 | 79 | 88 | 78 | lightning in W at 8 |
| 26 | Sun. | 4:57 | 7:03 | 75 | 89 | 81 | .36 last night |
| 27 | Mon. | 4:58 | 7:02 | 78 | 80 | 76 | light showers |
| 28 | Tues. | 4:59 | 7:01 | 78 | 88 | 77 | rain from 7 onward |
| 29 | Wed. | 4:59 | 7:01 | 76 | 82 | 75 | .14 |
| 30 | Thurs. | 5:00 | 7:00 | 71 | 86 | 77 | .49, rained all night |
| 31 | Fri. | 5:01 | 6:59 | 76 | 84 | 78 | |

August 1863

Confederates on the march near Orange Court House on August 1 endured what one of them called "the hottest sun that I ever felt. The men were constantly dropping out from overheat, and one or two died from the effects."[54]

August 2 in Richmond offered "warm, fair weather." A week later "hot August weather" prevailed. "A heavy shower" on the 29th afforded Richmonders relief from the heat and help for their gardens."[55]

A diarist camped near Gordonsville called August 6 and 7 "hot," but with "rain in evening."[56]

The Georgetown weather records for this month are missing—the only month-long lacuna during the entire war period. To cover the gap, the temperatures and rainfall remarks in this table for August 1863 come from a weather station at the U.S. General Hospital in Alexandria, Virginia. Comparing the Alexandria records to Georgetown's readings for other months shows that the Alexandria thermometer generally measured a little bit warmer. The Alexandria rain remarks appear in perfect alignment to the dates for August, so they can be relied upon for accurate reporting. The Alexandria observer(s) also figured the daily average for the three temperature reports, which provides a reassuring cross-checking device in instances of poorly written or obscured individual readings. For this uncommonly dry, clear, and hot month, Alexandria counted 26.33 fair days, only 2 cloudy days, and 2.67 rainy days.[57]

## August 1863

| Date | Day of Week | Rich-mond Sunrise | Rich-mond Sunset | Alexandria Temperature 7 a.m. | 2 p.m. | 9 p.m. | Other |
|------|------|------|------|------|------|------|------|
| 1 | Sat. | 5:02 | 6:58 | 86 | 92 | 84 | |
| 2 | Sun. | 5:03 | 6:57 | 80 | 98 | 86 | |
| 3 | Mon. | 5:04 | 6:56 | 80 | 98 | 85 | |
| 4 | Tues. | 5:05 | 6:55 | 85 | 100 | 81 | |
| 5 | Wed. | 5:06 | 6:54 | 88 | 90 | 83 | |
| 6 | Thurs. | 5:07 | 6:53 | 93 | 95 | 80 | |
| 7 | Fri. | 5:08 | 6:52 | 88 | 93 | 89 | rain ended 4 p.m. |
| 8 | Sat. | 5:09 | 6:51 | 87 | 94 | 79 | |
| 9 | Sun. | 5:10 | 6:50 | 88 | 98 | 85 | |
| 10 | Mon. | 5:11 | 6:49 | 93 | 104 | 80 | |
| 11 | Tues. | 5:12 | 6:48 | 94 | 101 | 88 | |
| 12 | Wed. | 5:13 | 6:47 | 84 | 90 | 78 | rained 3:30–5:30 p.m. |
| 13 | Thurs. | 5:14 | 6:46 | 76 | 80 | 77 | rained 7–9 a.m. |
| 14 | Fri. | 5:15 | 6:45 | 83 | 94 | 80 | |
| 15 | Sat. | 5:16 | 6:44 | 83 | 91 | 83 | |
| 16 | Sun. | 5:17 | 6:43 | 84 | 96 | 84 | |
| 17 | Mon. | 5:18 | 6:42 | 81 | 93 | 71 | |
| 18 | Tues. | 5:19 | 6:41 | 70 | 86 | 75 | |
| 19 | Wed. | 5:20 | 6:40 | 74 | 86 | 70 | |
| 20 | Thurs. | 5:21 | 6:39 | 66 | 91 | 75 | |
| 21 | Fri. | 5:22 | 6:38 | 77 | 93 | 79 | |
| 22 | Sat. | 5:23 | 6:37 | 75 | 93 | 88 | |
| 23 | Sun. | 5:24 | 6:36 | 86 | 99 | 83 | |
| 24 | Mon. | 5:25 | 6:35 | 81 | 95 | 80 | rained 2–4 p.m. |
| 25 | Tues. | 5:27 | 6:33 | 79 | 97 | 69 | rained 4:30–11 p.m. |
| 26 | Wed. | 5:28 | 6:32 | 61 | 75 | 64 | |
| 27 | Thurs. | 5:29 | 6:31 | 67 | 73 | 61 | |
| 28 | Fri. | 5:30 | 6:30 | 70 | 74 | 70 | |
| 29 | Sat. | 5:31 | 6:29 | 70 | 75 | 65 | rained 4:45–5:15 p.m. |
| 30 | Sun. | 5:32 | 6:28 | 59 | 71 | 59 | |
| 31 | Mon. | 5:33 | 6:27 | 57 | 75 | 63 | |

## September 1863

A government employee from Louisiana, living in Richmond, remarked in his diary on September 6 about the hospital he had just left. His comments reveal much about the Confederate capital's wartime environment: "It was extraordinary how many flies there were at the hospital." Two days later he noted, "We have not had any rain for several weeks. There is dust everywhere."[58]

On September 15 the weather prompted an artillerist in Richmond's defenses to write lyrically: "To-day is one of the brightest and sunniest of this bright and sweet autumnal month. The woods have put on their 'sear and yellow robe,' but the days are warm and splendid, the sky, intensely blue."[59]

An early cold snap brought "a pretty smart frost" near Rapidan Station on the night of September 20–21, "and the wind blew like winter." A North Carolinian camped there "spent two thirds of the night by the fire to keep warm."[60] An Alabama soldier who had grumbled about August's dreadful heat wrote on the morning of September 23, "killing frost this morning."[61]

One hundred miles north of Richmond, at Virginia's boundary on the Potomac, the drought broke a bit earlier than in the state capital, and an extensive wet spell began. The Alexandria weather station reported twenty-three fair days during the month, but rain on September 6, 11, 12, 13, 16, and 18.[62]

September 1863

| Date | Day of Week | Rich- mond Sunrise | Rich- mond Sunset | D.C. Temperature | | | Other |
|---|---|---|---|---|---|---|---|
| | | | | 7 a.m. | 2 p.m. | 9 p.m. | |
| 1 | Tues. | 5:34 | 6:26 | 60 | 77 | 70 | |
| 2 | Wed. | 5:35 | 6:25 | 62 | 78 | 63 | |
| 3 | Thurs. | 5:36 | 6:24 | 57 | 79 | 73 | |
| 4 | Fri. | 5:38 | 6:22 | 62 | 74 | 67 | |
| 5 | Sat. | 5:39 | 6:21 | 65 | 78 | 68 | |
| 6 | Sun. | 5:40 | 6:20 | 64 | 85 | 70 | slight shower at 4:30 p.m. |
| 7 | Mon. | 5:41 | 6:19 | 66 | 82 | 78 | |
| 8 | Tues. | 5:42 | 6:18 | 72 | 88 | 80 | |
| 9 | Wed. | 5:44 | 6:16 | 72 | 80 | 66 | dim horizon at 9:30 |
| 10 | Thurs. | 5:45 | 6:15 | 66 | 74 | 69 | |
| 11 | Fri. | 5:46 | 6:14 | 63 | 76 | 74 | |
| 12 | Sat. | 5:47 | 6:13 | 65 | 82 | 72 | high wind & shower 4 p.m. |
| 13 | Sun. | 5:48 | 6:12 | 68 | 81 | 76 | .48 yesterday & last night |
| 14 | Mon. | 5:49 | 6:11 | 68 | 78 | 73 | |
| 15 | Tues. | 5:50 | 6:10 | 70 | 85 | 78 | misting at 7 a.m. |
| 16 | Wed. | 5:51 | 6:09 | 73 | 79 | 75 | showers all day |
| 17 | Thurs. | 5:53 | 6:07 | 71 | 83 | 76 | light fog 7 a.m. |
| 18 | Fri. | 5:54 | 6:06 | 76 | 75 | 63 | 1.03 |
| 19 | Sat. | 5:55 | 6:05 | 58 | 64 | 57 | |
| 20 | Sun. | 5:56 | 6:04 | 54 | 60 | 50 | |
| 21 | Mon. | 5:58 | 6:02 | 45 | 70 | 54 | |
| 22 | Tues. | 5:59 | 6:01 | 5- | 64 | 49 | |
| 23 | Wed. | 6:00 | 6:00 | 43 | 66 | 53 | Autumnal Equinox |
| 24 | Thurs. | 6:01 | 5:59 | 50 | 72 | 59 | |
| 25 | Fri. | 6:03 | 5:57 | 53 | 66 | 52 | |
| 26 | Sat. | 6:04 | 5:56 | 46 | 62 | 47 | |
| 27 | Sun. | 6:05 | 5:55 | 51 | 65 | 49 | |
| 28 | Mon. | 6:07 | 5:53 | 44 | 69 | 52 | |
| 29 | Tues. | 6:08 | 5:52 | 43 | 72 | 52 | |
| 30 | Wed. | 6:09 | 5:51 | 50 | 74 | 58 | |

October 1863

Near Morton's Ford on the Rapidan River, the night of October 3–4 "was a terrible night, cold and rainy, and the wind was pretty cutting."[63]

On the 15th, as Virginia horsemen rode toward Bristoe Station, "it was raining hard." The next day they moved along the Manassas Gap Railroad "in a furious storm. There seemed to be nothing but water everywhere."[64]

Again on October 19, Lee's men campaigning in northern Virginia faced bad weather: "a cold, cheerless rain was falling with sullen determination, as if it would never cease. Rarely have I suffered more. . . . The cold rain trickled down my back and finally stopped in my boots, until even they were half filled with water."[65]

"A cold northwest storm of wind and rain" moved through Richmond on October 24, prompting householders to light their first fires of the autumn. Two days later the weather remained "cloudy and cold."[66]

Autumn weather in the Shenandoah Valley brought "bright but cold" skies to Front Royal on October 25. The 28th also was "very cold and clear" there.[67]

## October 1863

| Date | Day of Week | Richmond Sunrise | Richmond Sunset | D.C. Temperature | | | Other |
|---|---|---|---|---|---|---|---|
| | | | | 7 a.m. | 2 p.m. | 9 p.m. | |
| 1 | Thurs. | 6:10 | 5:50 | 56 | -8 | 62 | |
| 2 | Fri. | 6:11 | 5:49 | 63 | 68 | 63 | |
| 3 | Sat. | 6:12 | 5:48 | 66 | 71 | 59 | 1.42 last night & yesterday |
| 4 | Sun. | 6:13 | 5:47 | 59 | 70 | 60 | |
| 5 | Mon. | 6:14 | 5:46 | 49 | 60 | 45 | |
| 6 | Tues. | 6:16 | 5:44 | 43 | 62 | 53 | |
| 7 | Wed. | 6:17 | 5:43 | 54 | 66 | 66 | |
| 8 | Thurs. | 6:18 | 5:42 | 58 | 60 | 50 | |
| 9 | Fri. | 6:19 | 5:41 | 42 | 67 | 52 | |
| 10 | Sat. | 6:21 | 5:39 | 50 | 66 | 56 | |
| 11 | Sun. | 6:22 | 5:38 | 50 | 62 | 52 | |
| 12 | Mon. | 6:23 | 5:37 | 40 | 60 | 46 | |
| 13 | Tues. | 6:24 | 5:36 | 42 | 64 | 50 | |
| 14 | Wed. | 6:26 | 5:34 | 45 | 65 | 54 | |
| 15 | Thurs. | 6:27 | 5:33 | 59 | 66 | 64 | |
| 16 | Fri. | 6:28 | 5:32 | 64 | 71 | 64 | .47, at 2 raining heavily |
| 17 | Sat. | 6:29 | 5:31 | 54 | 77 | 62 | |
| 18 | Sun. | 6:30 | 5:30 | 64 | 78 | 64 | |
| 19 | Mon. | 6:31 | 5:29 | 51 | 65 | 53 | |
| 20 | Tues. | 6:32 | 5:28 | 49 | 67 | 50 | |
| 21 | Wed. | 6:33 | 5:27 | 49 | 75 | 60 | |
| 22 | Thurs. | 6:34 | 5:26 | 52 | 77 | 48 | |
| 23 | Fri. | 6:36 | 5:24 | 49 | 64 | 54 | |
| 24 | Sat. | 6:37 | 5:23 | 49 | 46 | 48 | |
| 25 | Sun. | 6:38 | 5:22 | 42 | 54 | 49 | |
| 26 | Mon. | 6:39 | 5:21 | 37 | 64 | 40 | |
| 27 | Tues. | 6:40 | 5:20 | 36 | 50 | 42 | |
| 28 | Wed. | 6:41 | 5:19 | 34 | 58 | 43 | |
| 29 | Thurs. | 6:43 | 5:17 | 34 | 55 | 46 | |
| 30 | Fri. | 6:44 | 5:16 | 46 | 66 | 60 | |
| 31 | Sat. | 6:45 | 5:15 | 58 | 62 | 46 | |

November 1863

November 1 developed into what a Richmonder called a "beautiful Sabbath day!" The next Sunday, by contrast, was "a bleak November day, after some days of pleasant autumnal sunshine." A front made November 9 "a dark and gloomy day, spitting snow," in Richmond.[68]

Mounted artillerists riding between Sperryville and Culpeper on November 9 encountered "the first snow of the season . . . and it was very cold." On the 16th, "it had been raining constantly for several days and the weather was very disagreeable."[69]

A journalist filing a telegraphic report from Lee's army at 11 a.m. on November 29 concluded that "the rain yesterday doubtless interfered with the fighting. It is cloudy this morning, but not rainy." Another reporter described the weather during this flurry of campaigning: "During Friday night [the 27th] a rain storm began, which lasted until late in the day on Saturday."[70]

As the armies faced off across Mine Run on November 30, the weather turned "very clear and cold, but we were glad to see the sun shining again, as we were worn out with so much wet and cold weather."[71]

November 1863

| Date | Day of Week | Rich-mond Sunrise | Rich-mond Sunset | D.C. Temperature | | | Other |
|---|---|---|---|---|---|---|---|
| | | | | 7 a.m. | 2 p.m. | 9 p.m. | |
| 1 | Sun. | 6:46 | 5:14 | 41 | 68 | 46 | |
| 2 | Mon. | 6:47 | 5:13 | 34 | 58 | 50 | |
| 3 | Tues. | 6:48 | 5:12 | 50 | 73 | 60 | |
| 4 | Wed. | 6:49 | 5:11 | 44 | 64 | 49 | |
| 5 | Thurs. | 6:50 | 5:10 | 52 | 74 | 67 | |
| 6 | Fri. | 6:51 | 5:09 | 54 | 62 | 48 | |
| 7 | Sat. | 6:52 | 5:08 | 36 | 62 | 54 | |
| 8 | Sun. | 6:53 | 5:07 | 52 | 57 | 46 | |
| 9 | Mon. | 6:54 | 5:06 | 37 | 46 | 37 | |
| 10 | Tues. | 6:55 | 5:05 | 31 | 43 | 38 | |
| 11 | Wed. | 6:56 | 5:04 | 30 | — | 50 | |
| 12 | Thurs. | 6:57 | 5:03 | 42 | 72 | 52 | |
| 13 | Fri. | 6:58 | 5:02 | 38 | 68 | 51 | |
| 14 | Sat. | 6:59 | 5:01 | 38 | 67 | 61 | |
| 15 | Sun. | 7:00 | 5:00 | 49 | 53 | 50 | 1.11, rained all night |
| 16 | Mon. | 7:01 | 4:59 | 46 | 57 | 48 | |
| 17 | Tues. | 7:01 | 4:59 | 40 | — | 45 | |
| 18 | Wed. | 7:02 | 4:58 | 46 | 57 | 46 | |
| 19 | Thurs. | 7:03 | 4:57 | 42 | 65 | 54 | |
| 20 | Fri. | 7:04 | 4:56 | 46 | 72 | 60 | |
| 21 | Sat. | 7:05 | 4:55 | 50 | 53 | 50 | .68, rained all day |
| 22 | Sun. | 7:06 | 4:54 | 46 | 55 | 46 | |
| 23 | Mon. | 7:07 | 4:53 | 34 | 59 | 48 | |
| 24 | Tues. | 7:07 | 4:53 | 45 | 50 | 52 | |
| 25 | Wed. | 7:08 | 4:52 | 50 | 52 | 43 | |
| 26 | Thurs. | 7:09 | 4:51 | 33 | 44 | 35 | |
| 27 | Fri. | 7:09 | 4:51 | 28 | — | 39 | |
| 28 | Sat. | 7:10 | 4:50 | 45 | 50 | 52 | |
| 29 | Sun. | 7:11 | 4:49 | 44 | 46 | 38 | |
| 30 | Mon. | 7:11 | 4:49 | 27 | 30 | 22 | |

## December 1863

December 1 dawned "bitter cold" at Mine Run. Because of the icy temperatures and "a gale of wind . . . everybody suffered terribly," a Federal colonel wrote in his diary. "There were rumours among the men of several of our pickets being frozen to death" during that frigid night, but the corps surgeon assured the colonel that the tales were exaggerated, "a few severe cases of frozen feet and hands being the worst."[72] Characteristically, the contemporary mythical tales continue to appear to this day, elaborated by time and often enshrined as demonstrable truth.

Reporting that "the weather has been bitter cold since Monday [November 30]," a Richmond newspaper issued a plea for blankets from families and the home front to keep Lee's men from freezing. A telegram from Mine Run to Richmond on December 5 offered better weather news: "The weather for the past two days has been most delightful."[73]

Confederates returning to their camps after Mine Run marched over "very muddy roads and in the coldest kind of weather." The 20th of the month was even worse—"the very coldest and most disagreeable we had experienced," wrote a Virginia gunner who kept close track of the weather.[74]

During December, the station at Alexandria observed just four days of rain and none with snow.[75]

The third Christmas of the war in Richmond seemed "sad," as shortages combined with weather that was "cold, and threatening snow" to dampen the season. Rain ("better than snow," a diarist thought) fell on December 27 and all day on New Year's Eve.[76]

A young recruit just leaving home in Bennettsville, South Carolina, reported that "a good snow" fell in that southerly latitude on Christmas Eve.[77]

December 1863

| Date | Day of Week | Rich- mond Sunrise | Rich- mond Sunset | D.C. Temperature | | | Other |
|---|---|---|---|---|---|---|---|
| | | | | 7 a.m. | 2 p.m. | 9 p.m. | |
| 1 | Tues. | 7:12 | 4:48 | 21 | 40 | 33 | |
| 2 | Wed. | 7:12 | 4:48 | 34 | 55 | 52 | |
| 3 | Thurs. | 7:13 | 4:47 | 39 | 60 | 40 | |
| 4 | Fri. | 7:13 | 4:47 | 36 | 61 | 50 | heavy white frost |
| 5 | Sat. | 7:14 | 4:46 | 34 | 53 | 41 | |
| 6 | Sun. | 7:14 | 4:46 | 32 | 40 | 30 | |
| 7 | Mon. | 7:15 | 4:45 | 20 | 38 | 28 | heavy white frost |
| 8 | Tues. | 7:15 | 4:45 | 18 | 42 | 30 | |
| 9 | Wed. | 7:15 | 4:45 | 28 | 52 | 38 | |
| 10 | Thurs. | 7:16 | 4:44 | 32 | 40 | 28 | |
| 11 | Fri. | 7:16 | 4:44 | 32 | 38 | 37 | |
| 12 | Sat. | 7:16 | 4:44 | 38 | 51 | 47 | drizzled at 4 p.m. |
| 13 | Sun. | 7:17 | 4:43 | 57 | 72 | 53 | drizzly at 7 a.m. |
| 14 | Mon. | 7:17 | 4:43 | 58 | 60 | 42 | |
| 15 | Tues. | 7:17 | 4:43 | 36 | 46 | 36 | |
| 16 | Wed. | 7:17 | 4:43 | 29 | 49 | 34 | |
| 17 | Thurs. | 7:18 | 4:42 | 34 | 42 | 42 | |
| 18 | Fri. | 7:18 | 4:42 | 40 | 42 | 34 | |
| 19 | Sat. | 7:18 | 4:42 | 30 | 31 | 24 | |
| 20 | Sun. | 7:18 | 4:42 | 22 | 30 | 24 | |
| 21 | Mon. | 7:18 | 4:42 | 18 | 33 | 32 | Winter Solstice |
| 22 | Tues. | 7:18 | 4:42 | 32 | 35 | 31 | |
| 23 | Wed. | 7:18 | 4:42 | 26 | 27 | 22 | |
| 24 | Thurs. | 7:18 | 4:42 | 19 | 36 | 27 | |
| 25 | Fri. | 7:18 | 4:42 | 22 | 34 | 28 | |
| 26 | Sat. | 7:18 | 4:42 | 15 | 38 | 32 | |
| 27 | Sun. | 7:17 | 4:43 | 37 | 40 | 30 | |
| 28 | Mon. | 7:17 | 4:43 | 38 | 44 | 38 | |
| 29 | Tues. | 7:17 | 4:43 | 32 | 44 | 34 | |
| 30 | Wed. | 7:17 | 4:43 | 32 | 58 | 36 | |
| 31 | Thurs. | 7:17 | 4:43 | 38 | 46 | 46 | drizzled all day |

# Chapter 5

# 1864

January 1864

Richmond opened the new year with "a bright windy day, and not cold." The temperature plummeted late on January 1, and the night became "bitter." On January 2 the morning dawned "bright and clear, and moderating." After a "dark and threatening" 4th, the 5th proved to be a "bright, pleasant day." When "a light snow" fell on Richmond on the 7th it was, amazingly, "the first time the earth has been white this winter." More snow overnight to the morning of the 8th cleared off into a "bitter cold" day.[1]

Lee's troops near the Rapidan River suffered under two snows during the week ending January 9, "each one two or three inches deep," driven by "cold north winds."[2]

In his tent in Culpeper County, a Pennsylvanian wrote in his diary on January 18: "Commenced raining about 6 A.M. and continued all day very heavy rain at noon."[3]

A Mississippi colonel called this winter, far north of his customary latitudes, "long, weary and vigorous."[4]

The Alexandria weather station reported these temperatures at 2 p.m. on days when the Georgetown recorder missed readings or left them illegible:

20 degrees on the 2nd, 40 on the 16th, and 40 on the 30th. Through January the weather in Alexandria featured many clouds but not much precipitation—twenty-one cloudy days but only one rainy day (the 18th), and snow on the 4th, 5th, and 7th.[5]

## January 1864

| Date | Day of Week | Rich-mond Sunrise | Rich-mond Sunset | D.C. Temperature | | | Other |
|---|---|---|---|---|---|---|---|
| | | | | 7 a.m. | 2 p.m. | 9 p.m. | |
| 1 | Fri. | 7:16 | 4:44 | 45 | 30 | 12 | |
| 2 | Sat. | 7:16 | 4:44 | 6 | — | 16 | |
| 3 | Sun. | 7:15 | 4:45 | 21 | 31 | 31 | |
| 4 | Mon. | 7:15 | 4:45 | 31 | 32 | 29 | snow 4 inches, began 10 a.m. |
| 5 | Tues. | 7:15 | 4:45 | 28 | 34 | 24 | |
| 6 | Wed. | 7:14 | 4:46 | 26 | 27 | — | |
| 7 | Thurs. | 7:14 | 4:46 | 9 | 25 | 22 | |
| 8 | Fri. | 7:13 | 4:47 | 16 | 28 | 22 | snow 1 inch last night |
| 9 | Sat. | 7:12 | 4:48 | 12 | 27 | -6 | |
| 10 | Sun. | 7:12 | 4:48 | 10 | 36 | 24 | |
| 11 | Mon. | 7:11 | 4:49 | 12 | 38 | 22 | |
| 12 | Tues. | 7:11 | 4:49 | 14 | 35 | 24 | |
| 13 | Wed. | 7:10 | 4:50 | — | 39 | 31 | |
| 14 | Thurs. | 7:09 | 4:51 | 22 | 40 | 37 | |
| 15 | Fri. | 7:09 | 4:51 | 34 | 47 | 40 | |
| 16 | Sat. | 7:08 | 4:52 | 27 | — | 26 | |
| 17 | Sun. | 7:07 | 4:53 | 31 | 44 | 37 | |
| 18 | Mon. | 7:06 | 4:54 | 38 | 42 | 38 | rain slight all night |
| 19 | Tues. | 7:06 | 4:54 | 44 | 45 | 34 | |
| 20 | Wed. | 7:05 | 4:55 | 31 | 43 | 36 | |
| 21 | Thurs. | 7:04 | 4:56 | 34 | 42 | 36 | |
| 22 | Fri. | 7:03 | 4:57 | 35 | 48 | 39 | |
| 23 | Sat. | 7:03 | 4:57 | 28 | 52 | — | |
| 24 | Sun. | 7:02 | 4:58 | 45 | 62 | 54 | |
| 25 | Mon. | 7:01 | 4:59 | 43 | 67 | 44 | |
| 26 | Tues. | 7:00 | 5:00 | 51 | 61 | 46 | |
| 27 | Wed. | 6:59 | 5:01 | 32 | 63 | 50 | |
| 28 | Thurs. | 6:58 | 5:02 | 36 | 72 | 56 | |
| 29 | Fri. | 6:57 | 5:03 | 36 | 68 | 52 | |
| 30 | Sat. | 6:56 | 5:04 | 42 | — | 40 | |
| 31 | Sun. | 6:55 | 5:05 | 40 | 48 | 42 | |

February 1864

"Hazy, misty weather" covered Richmond on February 1, but the next day became "beautiful and spring-like." Both the 3rd and the 4th were clear and cold. A snowstorm on the 15th melted completely during a "bright windy" 16th. February 17 in Richmond remained "very cold—freezing all day," but the 18th outdid it with "the coldest morning of the winter. There was ice in the wash-basins in our bed chambers, the first we have seen there." The only February 29 that the Confederate nation experienced brought moderate rain to Richmond.[6]

In describing this season, a diarist encamped near Charlottesville said, without specifying individual dates, that "a great deal of snow fell this winter, and altogether it was the severest in our experience."[7]

Near Paris, "a squall of snow" struck on the morning of February 16, ushering in "a severe cold day." Three days later a girl wrote in her diary there of "the coldest weather I have ever felt," and described visiting soldiers "frolicking over the ice pond," some of them having "disgrace[d] themselves with King Alcohol."[8]

A weather station in Alexandria, just across the Potomac River from Georgetown, recorded 2 p.m. readings of 50 degrees on the 8th and 30 degrees on the 10th. The Alexandria ledger showed twelve fair and seventeen cloudy days, with only one measurable snow.[9]

## February 1864

| Date | Day of Week | Rich-mond Sunrise | Rich-mond Sunset | D.C. Temperature | | | Other |
|------|-------------|-------------------|------------------|-------|--------|--------|-------|
| | | | | 7 a.m. | 2 p.m. | 9 p.m. | |
| 1 | Mon. | 6:54 | 5:06 | 42 | 48 | 48 | |
| 2 | Tues. | 6:53 | 5:07 | 38 | 51 | 44 | |
| 3 | Wed. | 6:52 | 5:08 | 34 | 47 | 32 | |
| 4 | Thurs. | 6:51 | 5:09 | 26 | 44 | 40 | |
| 5 | Fri. | 6:50 | 5:10 | 32 | 55 | 44 | |
| 6 | Sat. | 6:49 | 5:11 | 38 | 50 | 43 | |
| 7 | Sun. | 6:48 | 5:12 | 42 | 47 | 40 | |
| 8 | Mon. | 6:47 | 5:13 | 33 | — | 54 | |
| 9 | Tues. | 6:46 | 5:14 | 26 | 43 | 30 | |
| 10 | Wed. | 6:45 | 5:15 | 21 | — | 24 | |
| 11 | Thurs. | 6:44 | 5:16 | 17 | 38 | 32 | |
| 12 | Fri. | 6:43 | 5:17 | 32 | 54 | 40 | |
| 13 | Sat. | 6:42 | 5:18 | 32 | 55 | 40 | |
| 14 | Sun. | 6:41 | 5:19 | 43 | 53 | 39 | |
| 15 | Mon. | 6:40 | 5:20 | 31 | 38 | 33 | began to snow 3 p.m. |
| 16 | Tues. | 6:39 | 5:21 | 32 | 39 | 18 | |
| 17 | Wed. | 6:37 | 5:23 | 7 | 12 | 6 | |
| 18 | Thurs. | 6:36 | 5:24 | 6 | 12 | 12 | |
| 19 | Fri. | 6:35 | 5:25 | 6 | 20 | 18 | |
| 20 | Sat. | 6:34 | 5:26 | 14 | 42 | 35 | |
| 21 | Sun. | 6:32 | 5:28 | 30 | 48 | 40 | |
| 22 | Mon. | 6:31 | 5:29 | 30 | 48 | 44 | |
| 23 | Tues. | 6:30 | 5:30 | 32 | 66 | 52 | |
| 24 | Wed. | 6:29 | 5:31 | 46 | 56 | 43 | |
| 25 | Thurs. | 6:28 | 5:32 | 33 | 56 | 48 | |
| 26 | Fri. | 6:27 | 5:33 | 40 | 35 | 32 | very windy all day |
| 27 | Sat. | 6:26 | 5:34 | 30 | 53 | 44 | |
| 28 | Sun. | 6:25 | 5:35 | 44 | 67 | 56 | |
| 29 | Mon. | 6:24 | 5:36 | 50 | 51 | 42 | |

March 1864

As the infamous Dahlgren raid approached Richmond on March 1, "dark and raining" weather impeded its progress and aided the capital's defenders. The next morning revealed "a slight snow on the ground" from overnight, but "bright and cool" weather came in during the ensuing day and "bright and frosty" skies on the 3rd. March 4 also began "bright and frosty," but turned "warm and cloudy in the afternoon." The capital's weather followed that pattern through March 7—"bright and frosty" mornings, later "warm and pleasant." Typically unpredictable March weather developed at the end of the month: windy and warm on the 27th; "April-like" on the 28th; "a furious gale, eastern, and rain" on the 29th; rain all night on the 30th with "the wind blowing a gale from the east"; and "cloudy and cold" on the 31st.[10]

A Virginia heavy artillerist sent in hurried pursuit of Dahlgren on "a very rainy night and . . . not very warm" found that his shoddy government-issue shoes disintegrated in the messy weather, leaving him virtually barefoot: "The slush and mud . . . soon caused them to come to pieces. The soles were gone, the uppers flapping about my ankles, but my feet in the mud."[11]

New Yorkers reconnoitering near Madison Court House at the beginning of March, as a cover for Custer's raid toward Charlottesville, reported that steady rains had left that region "in many places covered with water."[12]

When a Tennessee soldier reached Orange Court House on March 14, during "quite a cold night," he took a room at a rickety hotel not far from the depot. He awakened the next morning to find three inches of snow on his bed, the snow having sifted in through the clapboards, and "about ten inches of snow" on the ground outside.[13]

A Virginia battery ordered from Charlottesville to Gordonsville on March 19 found their progress impeded by "muddy roads. . . . The weather was cold, with much snow and rain."[14]

Troops in winter quarters in Louisa County coped with "snow 8 inches deep" on March 22 and 23.[15]

## March 1864

| Date | Day of Week | Rich-mond Sunrise | Rich-mond Sunset | D.C. Temperature | | | Other |
|------|------|------|------|------|------|------|------|
| | | | | 7 a.m. | 2 p.m. | 9 p.m. | |
| 1 | Tues. | 6:23 | 5:37 | 33 | 33 | 33 | |
| 2 | Wed. | 6:22 | 5:38 | 24 | 44 | 35 | |
| 3 | Thurs. | 6:21 | 5:39 | 28 | 48 | 42 | |
| 4 | Fri. | 6:20 | 5:40 | 34 | 64 | 48 | |
| 5 | Sat. | 6:19 | 5:41 | 46 | 51 | 40 | |
| 6 | Sun. | 6:17 | 5:43 | 39 | 44 | 39 | |
| 7 | Mon. | 6:16 | 5:44 | 34 | 53 | 42 | |
| 8 | Tues. | 6:15 | 5:45 | 39 | 44 | 41 | |
| 9 | Wed. | 6:14 | 5:46 | 30 | 57 | 50 | |
| 10 | Thurs. | 6:12 | 5:48 | 44 | 44 | 47 | |
| 11 | Fri. | 6:11 | 5:49 | 44 | 46 | 46 | |
| 12 | Sat. | 6:10 | 5:50 | 38 | 62 | 48 | |
| 13 | Sun. | 6:09 | 5:51 | 40 | 61 | 50 | |
| 14 | Mon. | 6:08 | 5:52 | 40 | 48 | 42 | |
| 15 | Tues. | 6:07 | 5:53 | 36 | 42 | 32 | |
| 16 | Wed. | 6:06 | 5:54 | 24 | 34 | 30 | |
| 17 | Thurs. | 6:05 | 5:55 | 30 | 45 | 36 | |
| 18 | Fri. | 6:03 | 5:57 | 34 | 58 | 52 | |
| 19 | Sat. | 6:02 | 5:58 | 38 | 54 | 42 | |
| 20 | Sun. | 6:00 | 6:00 | 32 | 50 | 33 | Vernal Equinox |
| 21 | Mon. | 5:59 | 6:01 | 24 | 36 | 32 | |
| 22 | Tues. | 5:58 | 6:02 | 25 | 31 | 26 | |
| 23 | Wed. | 5:57 | 6:03 | 24 | 40 | 33 | |
| 24 | Thurs. | 5:55 | 6:05 | 30 | 50 | 38 | |
| 25 | Fri. | 5:54 | 6:06 | 33 | 44 | 37 | |
| 26 | Sat. | 5:53 | 6:07 | 40 | 55 | 44 | |
| 27 | Sun. | 5:52 | 6:08 | 44 | 58 | 47 | Easter Sunday |
| 28 | Mon. | 5:51 | 6:09 | 37 | 64 | 48 | |
| 29 | Tues. | 5:50 | 6:10 | 43 | 48 | 45 | |
| 30 | Wed. | 5:49 | 6:11 | 34 | — | 38 | |
| 31 | Thurs. | 5:48 | 6:12 | 42 | 52 | 45 | |

April 1864

"Large snowflakes" fell on Richmond early on April 2 after a night-long rain, making for "abominable" weather. As the morning warmed, the snow changed to rain and came down all day long. Again on the 4th, "it rained all day and was cold." The rain persisted through that night and all of the 5th too, clearing finally only during the afternoon of the 6th.[16]

A temporary weather station at the Federal supply point "Rendezvous Distribution, Va.," situated about ten miles northeast of Warrenton, provides readings for 7 a.m., 2 p.m., and 9 p.m. on April 20 and 21, to make up for the disfigured portion of the Georgetown sheet: 44, 54, 45; and 54, 56, 49.[17]

After a wet spell at mid-month, Richmond's weather remained "cloudy and cold" on April 19 and 20, then changed to "bright sunshine all day, but cool," on the 21st. Cherry trees in the city burst into blossom on the 22nd, "a bright day and warmer." The next several days remained warm and generally sunny: "A bright day, with southern breezes" on the 23rd; "cloudy and windy, but warm" on the 24th; "bright and beautiful" on the 25th and 27th; "another truly fine spring day" on the 26th.[18]

General Lee's troops along the Rapidan River line enjoyed the arrival of spring. On April 21 a North Carolinian wrote of the week past as "the most pleasant we have had this winter. Only one day and night of rain, the rest of the time the most delightful kind of weather."[19]

On April 29, Colonel James Conner (destined to become a general within a few weeks) told his mother in a letter from the Rapidan River camps: "The weather here is delightful, peach and apple trees in full blossom."[20]

## April 1864

| Date | Day of Week | Richmond Sunrise | Richmond Sunset | D.C. Temperature | | | Other |
|------|-------------|------------------|-----------------|--------|--------|--------|-------|
| | | | | 7 a.m. | 2 p.m. | 9 p.m. | |
| 1 | Fri. | 5:46 | 6:14 | 44 | 56 | 45 | |
| 2 | Sat. | 5:45 | 6:15 | 36 | 37 | 38 | .81, snowing & raining |
| 3 | Sun. | 5:44 | 6:16 | 41 | 56 | 52 | |
| 4 | Mon. | 5:43 | 6:17 | 37 | 42 | 39 | |
| 5 | Tues. | 5:41 | 6:19 | 37 | 40 | 46 | .81, windy & rain |
| 6 | Wed. | 5:40 | 6:20 | 40 | 58 | 48 | cleared off at noon |
| 7 | Thurs. | 5:39 | 6:21 | 40 | 65 | 50 | |
| 8 | Fri. | 5:38 | 6:22 | 42 | 70 | 54 | |
| 9 | Sat. | 5:36 | 6:24 | 50 | 48 | 46 | |
| 10 | Sun. | 5:35 | 6:25 | 56 | 68 | 50 | |
| 11 | Mon. | 5:34 | 6:26 | 48 | 60 | 50 | |
| 12 | Tues. | 5:33 | 6:27 | — | 66 | 51 | |
| 13 | Wed. | 5:32 | 6:28 | 46 | 60 | 47 | |
| 14 | Thurs. | 5:31 | 6:29 | 46 | 58 | 51 | |
| 15 | Fri. | 5:30 | 6:30 | 46 | 62 | 50 | |
| 16 | Sat. | 5:29 | 6:31 | 46 | 48 | 44 | |
| 17 | Sun. | 5:28 | 6:32 | 47 | 52 | 46 | |
| 18 | Mon. | 5:26 | 6:34 | 44 | 62 | 54 | |
| 19 | Tues. | 5:25 | 6:35 | 43 | 60 | 46 | |
| 20 | Wed. | 5:24 | 6:36 | — | — | — | |
| 21 | Thurs. | 5:23 | 6:37 | — | — | — | |
| 22 | Fri. | 5:22 | 6:38 | 45 | 68 | 54 | |
| 23 | Sat. | 5:21 | 6:39 | 58 | 78 | 68 | |
| 24 | Sun. | 5:20 | 6:40 | 68 | 77 | 65 | |
| 25 | Mon. | 5:19 | 6:41 | 65 | 76 | 61 | |
| 26 | Tues. | 5:18 | 6:42 | 57 | 68 | 56 | |
| 27 | Wed. | 5:17 | 6:43 | 57 | 70 | 63 | |
| 28 | Thurs. | 5:16 | 6:44 | 44 | 60 | 48 | at 6 p.m. quite a tornado of wind |
| 29 | Fri. | 5:15 | 6:45 | 41 | 65 | 54 | |
| 30 | Sat. | 5:14 | 6:46 | 63 | 64 | 57 | |

## May 1864

May 2 dawned "fair and lovely" near the military frontier along the Rapidan River, "but in the afternoon the skies were overspread with clouds and angry west winds came sweeping from the mountains with a hurricane-like sound . . . then came the rain and hail, driving full into our faces."[21]

As Grant moved into the Wilderness early in the month, central Virginia enjoyed gorgeous weather. A diarist in Richmond called May 4 "a magnificent day with bright sunshine" and warming temperature. He reported the 6th as "a glorious day with bright sunshine," but by the 7th it had turned "extremely hot."[22]

A Confederate artillerist marching toward what would become Wilderness battlefield called May 5 "very hot and dusty" in his diary.[23]

The day after the Wilderness, May 7, proved to be "very quiet and quite warm." The temperature sped decay in the bodies of dead Federals covering Saunders Field on the Orange Turnpike, which already had become "very much swollen."[24]

On May 11 and 12, while Lee's soldiers struggled desperately at Spotsylvania's Bloody Angle,[25] home front and local units protected Richmond against Sheridan's raiders. One of the defenders grumbled in his journal about intense dust on May 10, but then described "a very heavy rain" on the afternoon of the 11th. On the 12th, "rained off & on during the day," then "it rained mostly during the night." The 14th again was a "rainy day" in the trenches just north of Richmond.[26]

The log of the USS *Agawam,* operating on the James River, recorded rain on the evening of May 16; on May 18–19; beginning on May 24 "and last[ing] during portions of three days"; and again on the night of May 28.[27]

The temporary Federal weather station at "Rendezvous Station," a few miles northeast of Warrenton, reported carefully throughout May. It recorded 72 degrees at 2 p.m. on May 26, at a time when the Georgetown entry is obscured. For the month, the temporary station observed seventeen fair days, fourteen cloudy days, and five with rain. As the armies clashed in the Wilder-

The twenty-hour fight at hand-to-hand range over Spotsylvania's "Bloody Angle" on May 12, 1864, raged under a steady rain that filled the trenches with mud, water, and blood. (Robert Underwood Johnson and C. C. Buel, eds., *Battles and Leaders of the Civil War: Being for the Most Part Contributions by Union and Confederate Officers*, 4 vols. [New York: The Century Co., 1887–88], 4:172.)

ness, then moved on to the opening days at Spotsylvania, skies remained bright blue. The thrice-daily readings from May 4 to 10 noted the presence of any clouds on only three of twenty-one possible occasions. Rain affected Union attack preparations late on May 11 and added to the savage misery of men fighting endlessly at Spotsylvania's renowned "Bloody Angle" on May 12. The front hovering over Spotsylvania perhaps did not extend as heavily the few dozen miles north to Rendezvous, because precipitation there measured only .39 inches, beginning at 9 p.m. on the 11th and recurring during the 12th from 11 a.m. to 1 p.m. Rendezvous reported heavier rain on May 14–15, totaling 1.2 inches (.80 of that on the 15th). A Federal soldier who lay wounded in the Wilderness wrote in his diary that the May 12 rain began at 9 a.m. and fell "rapidly . . . it seems as if Heaven has to weep over the scene."[28]

Because of the proximity of the temporary Rendezvous station to the scenes of the Overland campaign, it seems worthwhile to report here the readings through the heart of the Wilderness and Spotsylvania fighting. Those daily readings cover the same three times used almost everywhere else, 7 a.m., 2 p.m., and 9 p.m.: 4th: 49, 67, 59; 5th: 52, 75, 60; 6th: 57, 91, 73; 7th: 63, 91, 74; 8th: 67, 95, 75; 9th: 72, 85, 75; 10th: 72, 93, 78; 11th: 71, 81, 62; 12th: 60, 68, 64; 13th: 63, 69, 63.[29]

The cavalry battle north of Richmond at Yellow Tavern began in hot and dusty weather on May 11, but by the time the mortally wounded General Jeb Stuart headed for Richmond in an ambulance, a violent spring storm had blown in. A Richmond newspaper on May 12 vividly described the weather and its consequences: "During the greater part of yesterday the weather was hot, dry and very windy. Clouds of dust filled the air and penetrated the eyes and noses of all militiamen who were abroad. But about six o'clock we were visited by a most welcome and refreshing shower, which laid the dust, washed out the streets, and made the trees and grass assume a livelier tint of green. During the gale that attended the rain last evening, the steeple of St. John's Church was blown down and fell in an easterly direction, smashing a number of tombstones. The back porch of Mr. Liggon's house, in the same neighborhood, was blown away, and . . . lodged on the top of the house." At St. John's, in the eastern Richmond neighborhood of Church Hill, Patrick Henry had famously declared "Give me Liberty or Give me Death" during an earlier revolution. The falling steeple "somewhat scaled off" two edges of the historic bell in the church tower, but it survived otherwise.[30]

The Confederate War Department clerk whose published diary became one of the most renowned sources on Richmond at war vividly described the storm of May 11: "I shall never forget the conformation of the clouds . . . as the storm arose. There were different strata running in various directions. They came in heaviest volume from the southeast in parallel lines, like lines of battle swooping over the city. There were at the same time shorter and fuller lines from the southwest, and others from the north. The meeting of these was followed by tremendous clashes of lightning and thunder; and between the pauses of the artillery of the elements above, the thunder of artillery on earth could be distinctly heard."[31]

A small heterogeneous Confederate force won the Battle of New Market on May 15 in the Shenandoah Valley. Youthful cadets from the Virginia Military Institute contributed to the triumph by launching a desperate, costly charge up a slope toward enemy artillery at the height of a violent rainstorm. "The rain was falling in torrents," one of them wrote, "and the smoke obscured our view—the noise was deafening." Deep mud in the open field through which they attacked sucked the shoes off some of the boys. The pouring rain held down the smoke from rifles and cannon, which "became so dense that I could only see the flash of their guns," in the words of an attacker. The success of the boys' attack became one of the most renowned small episodes of the war, and artistic portrayals of their triumph amidst a violent storm are perhaps the best-known depictions of dramatic Civil War weather.[32]

A Confederate soldier fighting near Drewry's Bluff described the morning of May 16: "The atmosphere was cool and bracing, but a dense fog prevailed, hiding everything from view fifty yards away."[33]

Along the route of Lee's clashes with Federals across an arc north of Richmond, mid-May weather conformed to customary patterns. A staff officer's diary recorded the weather at the front: May 17, "very warm," with an evening shower; 18th, "a very heavy shower at noon," but "quite pleasant"; 20th, "a very fine day"; 21st, "very pleasant and mild." He reported some rain on each day from May 24 to 28.[34]

Cadets of the Virginia Military Institute charged an enemy battery at New Market through a violent thunderstorm on the afternoon of May 15, 1864. (Virginia Military Institute)

Triumphant cadets of the Virginia Military Institute leaped atop a captured cannon at New Market on May 15, 1864, to celebrate an amazing victory won at the height of a noisy spring storm. (Virginia Military Institute)

May 1864

| Date | Day of Week | Rich-mond Sunrise | Rich-mond Sunset | D.C. Temperature | | | Other |
|------|-------------|-------------------|------------------|------------------|--------|--------|-------|
| | | | | 7 a.m. | 2 p.m. | 9 p.m. | |
| 1 | Sun. | 5:13 | 6:47 | 55 | 63 | 52 | |
| 2 | Mon. | 5:12 | 6:48 | 48 | 68 | 42 | |
| 3 | Tues. | 5:11 | 6:49 | 47 | 59 | 50 | |
| 4 | Wed. | 5:10 | 6:50 | 52 | 68 | 53 | |
| 5 | Thurs. | 5:09 | 6:51 | 54 | 70 | 62 | |
| 6 | Fri. | 5:08 | 6:52 | 58 | 80 | 68 | |
| 7 | Sat. | 5:07 | 6:53 | 68 | 89 | 72 | |
| 8 | Sun. | 5:06 | 6:54 | 78 | 90 | 72 | |
| 9 | Mon. | 5:05 | 6:55 | 70 | 86 | 75 | heavy thunder & brilliant lightning at 9 p.m. |
| 10 | Tues. | 5:04 | 6:56 | 69 | 89 | 80 | |
| 11 | Wed. | 5:03 | 6:57 | 70 | 84 | 60 | .24, thunder & lightning at 5:30 p.m. |
| 12 | Thurs. | 5:02 | 6:58 | 59 | 76 | 62 | at 4:30 heavy shower, considerable hail |
| 13 | Fri. | 5:01 | 6:59 | 62 | 72 | 64 | .74 yesterday and today |
| 14 | Sat. | 5:00 | 7:00 | 66 | 76 | 68 | |
| 15 | Sun. | 4:59 | 7:01 | 60 | 68 | 59 | heavy rain thru the day |
| 16 | Mon. | 4:58 | 7:02 | 66 | 76 | 64 | .45. At 9 p.m. clear moonlight. |
| 17 | Tues. | 4:57 | 7:03 | — | 76 | 65 | |
| 18 | Wed. | 4:57 | 7:03 | 66 | 70 | 64 | |
| 19 | Thurs. | 4:56 | 7:04 | 65 | 68 | 65 | |
| 20 | Fri. | 4:55 | 7:05 | 62 | 78 | 64 | |
| 21 | Sat. | 4:55 | 7:05 | 65 | 87 | 72 | |
| 22 | Sun. | 4:54 | 7:06 | 78 | 86 | 73 | |
| 23 | Mon. | 4:53 | 7:07 | 62 | 78 | 70 | |
| 24 | Tues. | 4:53 | 7:07 | 70 | 81 | 67 | |
| 25 | Wed. | 4:52 | 7:08 | 70 | 83 | 68 | .18 last night |
| 26 | Thurs. | 4:51 | 7:09 | 75 | — | 65 | raining, drizzle |
| 27 | Fri. | 4:51 | 7:09 | 63 | 79 | 66 | .64 |
| 28 | Sat. | 4:50 | 7:10 | 70 | 82 | 64 | |
| 29 | Sun. | 4:49 | 7:11 | 62 | 70 | 62 | |
| 30 | Mon. | 4:49 | 7:11 | 65 | 84 | 71 | |
| 31 | Tues. | 4:48 | 7:12 | 69 | 90 | 76 | |

## June 1864

As Federal armies approached Richmond to begin a campaign that would last for ten months, June dawned "lovely" and "cool" on the 1st, but by that afternoon it had become "very hot . . . in the sun." June 2 began "a bit cool," and again the "evening . . . was a little cool." The next day, as Lee's army shot down Northern troops in hecatombs, and again on the 4th, cool weather prevailed. So did "cursed flies," which "tormented" everyone mercilessly: "it is impossible to imagine the number of flies that we have here." Troops digging the trenches that eventually would encircle the city began to complain of the effect of oppressive heat, but in the city, cool mornings and evenings continued into mid-June.[35]

An Alabama soldier of General Evander M. Law's brigade, which had butchered Grant's unfortunate attackers with particularly good effect at Cold Harbor on June 3, described the time of the initial onset as "about twilight" on "a very foggy morning," but the day subsequently turned "very warm."[36]

A Union navy gunboat on the James River recorded some rain during the nights of June 2 and 3, and a full-fledged rainstorm that began in the afternoon of June 4 and lasted about twenty-four hours.[37]

In the trenches near Cold Harbor on June 5, Confederate defenders found "plenty of rain and no rest." Rain fell again all night on the 6th.[38]

Local defense troops repulsed Northern cavalry raiders who threatened Petersburg on June 9 in what came to be known as "The Battle of Old Men and Young Boys." One of the defenders wrote that the morning of their famous defense "dawned . . . with as fair a face and as bright a promise as the best . . . in all that queenly Spring."[39]

Richmond's weather remained cooler than the norm during mid-June: "Cold and cloudy" on the 12th; "clear and cool" on the 13th and 14th; the same on the 15th, but "warm late in the day"; "clear and pleasant" on the 17th; cool on the 18th and 19th, and "hazy" on the latter day. The capital city's climate inevitably turned warmer as the season advanced: "Hot and hazy; dry," on the 24th; "hot and dry" on the 25th and 26th, but "breezy" on the latter; "bright and hot—afterward light showers" on the 27th.[40]

A strong detachment sent toward the Shenandoah Valley by Lee saved Lynchburg from capture in mid-June. The Confederates found the weather there "very hot and dusty" on the 17th, "very warm" on the 18th, and "warm . . . dusty and dry" as they began to pursue the routed Federals on the 19th.[41]

A South Carolina cavalryman riding through dry and hot country near Richmond on June 24, headed for General Wade Hampton's battle at Nance's Shop, wrote: "The dust was ankle deep on the horses . . . the dust was so intense that we could not see the horses immediately in our front. . . . It came near suffocating us and I spit out great mouthfuls of pure mud."[42]

June 1864

| Date | Day of Week | Rich-mond Sunrise | Rich-mond Sunset | D.C. Temperature | | | Other |
|---|---|---|---|---|---|---|---|
| | | | | 7 a.m. | 2 p.m. | 9 p.m. | |
| 1 | Wed. | 4:48 | 7:12 | 71 | 92 | 77 | 2:00 sprinkling |
| 2 | Thurs. | 4:47 | 7:13 | 68 | 90 | 64 | occ. slight showers |
| 3 | Fri. | 4:47 | 7:13 | 64 | 81 | 64 | |
| 4 | Sat. | 4:46 | 7:14 | 59 | 68 | 70 | |
| 5 | Sun. | 4:46 | 7:14 | 68 | 76 | 68 | first fireflies this evening |
| 6 | Mon. | 4:45 | 7:15 | 70 | 90 | — | .60, heavy rain & high wind at 5 p.m. |
| 7 | Tues. | 4:45 | 7:15 | 64 | 74 | 63 | |
| 8 | Wed. | 4:44 | 7:16 | 63 | 76 | 70 | |
| 9 | Thurs. | 4:44 | 7:16 | 72 | 86 | — | |
| 10 | Fri. | 4:44 | 7:16 | 64 | 72 | 62 | |
| 11 | Sat. | 4:44 | 7:16 | 59 | 75 | 60 | |
| 12 | Sun. | 4:43 | 7:17 | 64 | 74 | 60 | |
| 13 | Mon. | 4:43 | 7:17 | 61 | 74 | 56 | |
| 14 | Tues. | 4:43 | 7:17 | 55 | 67 | 58 | |
| 15 | Wed. | 4:43 | 7:17 | 64 | 81 | 69 | |
| 16 | Thurs. | 4:42 | 7:18 | — | 82 | 72 | |
| 17 | Fri. | 4:42 | 7:18 | 72 | 91 | 77 | |
| 18 | Sat. | 4:42 | 7:18 | 68 | 88 | 68 | |
| 19 | Sun. | 4:42 | 7:18 | 65 | 84 | 70 | |
| 20 | Mon. | 4:42 | 7:18 | 64 | 88 | 66 | |
| 21 | Tues. | 4:42 | 7:18 | 63 | 86 | 66 | Summer Solstice |
| 22 | Wed. | 4:42 | 7:18 | 64 | 80 | 72 | |
| 23 | Thurs. | 4:42 | 7:18 | 74 | 92 | 82 | |
| 24 | Fri. | 4:42 | 7:18 | 8- | 95 | 82 | |
| 25 | Sat. | 4:42 | 7:18 | 81 | 98 | 85 | |
| 26 | Sun. | 4:42 | 7:18 | 86 | — | 84 | 94.5 at 10:30 a.m. |
| 27 | Mon. | 4:43 | 7:17 | 76 | 97 | 74 | very heavy wind 2 p.m. |
| 28 | Tues. | 4:43 | 7:17 | 64 | 82 | 68 | |
| 29 | Wed. | 4:43 | 7:17 | 65 | 83 | 70 | |
| 30 | Thurs. | 4:43 | 7:17 | 68 | 78 | 71 | smart shower at 3 p.m. |

# July 1864

July in the Confederate capital opened with typically oppressive mid-summer weather: "Clear, hot, and dry." Similar reports continued through the first ten days of the month, leading to concern expressed on the 10th: "The drought continues; vegetation wilting and drying up. . . . The gardens are nearly ruined." Relief appeared on the evening of the 11th when "a gentle shower, lasting more than an hour," fell at 7 p.m. Through the next week the drought reasserted its hold on central Virginia, then finally on the 19th, "A steady, gentle rain [fell] from 8 a.m. till 4 p.m."[43]

The same hot July weather wrapped the Shenandoah Valley in its clutches. A lieutenant from North Carolina recorded his observations in a diary: 1st, "hot"; 4th, "hot and dusty"; 7th, marching at night through "rain, rough and very dark. . . . awful rain . . . all and everything wet through."[44]

"Rained last night," a Virginia artillerist camped on the Darbytown Road outside Richmond wrote in his diary on July 25.[45]

In Richmond on July 30, as the Federal mine explosion near Petersburg triggered the Battle of the Crater, a civilian diarist complained that even at daybreak "it was very hot." He described "a burning sun" at 9 a.m., and by midafternoon the air had grown so "very hot" that he remained in his room, where a bath did nothing to assuage the intolerable heat.[46]

Confederates fighting in the Shenandoah Valley late in July encountered more typically seasonal weather. After a "very pleasant" July 25 and a "fine day" on the 26th, the month's last five days were "warm"—"very warm" for three days running, then "warm" on the 31st.[47]

## July 1864

| Date | Day of Week | Rich-mond Sunrise | Rich-mond Sunset | D.C. Temperature | | | Other |
|------|-------------|-------------------|------------------|------|------|------|-------|
| | | | | 7 a.m. | 2 p.m. | 9 p.m. | |
| 1 | Fri. | 4:43 | 7:17 | 72 | — | 80 | |
| 2 | Sat. | 4:44 | 7:16 | 80 | — | 77 | |
| 3 | Sun. | 4:44 | 7:16 | 78 | 88 | 78 | |
| 4 | Mon. | 4:44 | 7:16 | 72 | — | 66 | 81 at 4 p.m. |
| 5 | Tues. | 4:45 | 7:15 | 67 | — | 72 | |
| 6 | Wed. | 4:45 | 7:15 | 69 | 93 | 74 | |
| 7 | Thurs. | 4:45 | 7:15 | 74 | 94 | 80 | |
| 8 | Fri. | 4:46 | 7:14 | 84 | 98 | 80 | |
| 9 | Sat. | 4:46 | 7:14 | 76 | 92 | 74 | |
| 10 | Sun. | 4:47 | 7:13 | — | 92 | 82 | |
| 11 | Mon. | 4:47 | 7:13 | — | 90 | 79 | |
| 12 | Tues. | 4:48 | 7:12 | 78 | 90 | 76 | |
| 13 | Wed. | 4:48 | 7:12 | 75 | 90 | 74 | |
| 14 | Thurs. | 4:49 | 7:11 | 74 | 88 | 73 | |
| 15 | Fri. | 4:50 | 7:10 | 72 | 84 | 68 | |
| 16 | Sat. | 4:50 | 7:10 | 64 | 86 | 70 | |
| 17 | Sun. | 4:51 | 7:09 | 73 | 92 | 72 | |
| 18 | Mon. | 4:52 | 7:08 | 72 | 86 | 73 | |
| 19 | Tues. | 4:52 | 7:08 | 74 | 90 | 78 | |
| 20 | Wed. | 4:53 | 7:07 | 77 | 93 | 76 | very heavy wind 4:30 p.m. |
| 21 | Thurs. | 4:54 | 7:06 | 72 | 80 | 70 | |
| 22 | Fri. | 4:55 | 7:05 | 62 | 78 | 60 | |
| 23 | Sat. | 4:55 | 7:05 | 59 | 84 | 65 | |
| 24 | Sun. | 4:56 | 7:04 | 68 | 84 | 73 | |
| 25 | Mon. | 4:57 | 7:03 | 61 | 72 | 66 | |
| 26 | Tues. | 4:57 | 7:03 | 62 | — | 73 | |
| 27 | Wed. | 4:58 | 7:02 | 74 | 92 | 75 | |
| 28 | Thurs. | 4:59 | 7:01 | 77 | 94 | 80 | |
| 29 | Fri. | 4:59 | 7:01 | 77 | 98 | 82 | |
| 30 | Sat. | 5:00 | 7:00 | 81 | 97 | 84 | |
| 31 | Sun. | 5:01 | 6:59 | 80 | 98 | 84 | |

August 1864

The Georgetown weather recorder wrote dates on his second (recto) ledger page during this month, so his comments and record of rainfall can be matched with the days. Those features therefore appear in full detail for this month. Through much of the rest of the war, the uncertain alignment between pages complicates the usefulness of some months' second pages.

A Richmond diarist described the air as "already hot" at 6 a.m. on August 1, but "a good breeze" cooled the next morning at the same hour. On the 3rd, "weather was threatening all day, but we still had no rain." Repeatedly during the rest of the month he complained of intense heat and pervasive dust.[48]

Another Richmond resident who kept a diary said about the same thing on August 9 that he did on many other days during the month: "Very hot; very dry; very dusty." A mere fifteen minutes of rain on the 10th afforded false hope, then finally on the 12th "at 3 p.m. rained about three minutes. We are burning up." When it began "raining gently" at 5 p.m. on the 19th, the diarist exclaimed "thank Heaven!" The beneficial rain continued hard all night and well into the 20th, leaving the ground "saturated." More rain on the evenings of the 25th ("copiously") and 26th ("fine shower") helped to moderate the drought.[49]

In an action called variously Second Deep Bottom or Fussell's Mill, Federals marching and fighting on August 14 and 15 complained of "temperature . . . something dreadful." Marching men "literally passed between men lying on both sides, dead from sunstroke." One brigadier reported 105 men overcome by heat in just two regiments. "The rays of the August sun smote the heads of the weary soldiers with blows as palpable as if they had been given with a club."[50]

Cavalry fighting outside Richmond on August 16 claimed the life of General John R. Chambliss. A colonel in his brigade called this day "the most trying our regiment had ever experienced. Not one drop of water could be had; the heat was intense, and the wood was dense and tangled."[51]

In the notably cooler Shenandoah Valley, a soldier wrote on August 19: "It is raining and the weather is cold enough to sleep under blankets." He reported rain near Winchester again the next day and on the 22nd.[52]

During the fighting near Ream's Station, a Virginia cavalry officer wrote to his mother on August 27 about reaching Tabernacle Church at midnight "amidst the severest storm of thunder, rain, and lightning that I have witnessed during my life. The storm was terribly grand."[53]

In the Shenandoah Valley, where General Jubal A. Early faced strong enemy forces, a North Carolinian wrote to his mother from Bunker Hill on August 30: "The weather has turned some cooler, the nights are quite cool, making a heavy blanket feel quite comfortable."[54]

## August 1864

| Date | Day of Week | Rich-mond Sunrise | Rich-mond Sunset | D.C. Temperature 7 a.m. | 2 p.m. | 9 p.m. | Other |
|------|-------------|-------------------|------------------|--------------------------|--------|--------|-------|
| 1 | Mon. | 5:02 | 6:58 | 79 | 98 | 84 | |
| 2 | Tues. | 5:03 | 6:57 | — | — | 78 | slight shower 6 p.m. |
| 3 | Wed. | 5:04 | 6:56 | 76 | — | 73 | |
| 4 | Thurs. | 5:05 | 6:55 | 76 | 84 | 74 | |
| 5 | Fri. | 5:06 | 6:54 | 68 | 91 | 78 | |
| 6 | Sat. | 5:07 | 6:53 | 71 | 77 | 75 | .55, shower at 2 p.m. |
| 7 | Sun. | 5:08 | 6:52 | 80 | 91 | 80 | |
| 8 | Mon. | 5:09 | 6:51 | — | 93 | 78 | |
| 9 | Tues. | 5:10 | 6:50 | 76 | 94 | 78 | |
| 10 | Wed. | 5:11 | 6:49 | 78 | 95 | 84 | |
| 11 | Thurs. | 5:12 | 6:48 | 78 | 97 | 86 | windy & lightning 9 p.m. |
| 12 | Fri. | 5:13 | 6:47 | 76 | 96 | 82 | |
| 13 | Sat. | 5:14 | 6:46 | 76 | 96 | 78 | thunder, lightning, shower at 5 p.m. |
| 14 | Sun. | 5:15 | 6:45 | 82 | 92 | 80 | thunder at 2 p.m., heavy shower at 5:30 |
| 15 | Mon. | 5:16 | 6:44 | 76 | 90 | 78 | slight shower at 5, lightning at 8 p.m. |
| 16 | Tues. | 5:17 | 6:43 | 70 | 88 | 79 | |
| 17 | Wed. | 5:18 | 6:42 | 80 | 91 | 78 | .74, shower all day |
| 18 | Thurs. | 5:19 | 6:41 | 76 | 82 | 78 | |
| 19 | Fri. | 5:20 | 6:40 | 71 | 80 | 73 | |
| 20 | Sat. | 5:21 | 6:39 | 78 | 80 | 70 | .56, showery from 2 |
| 21 | Sun. | 5:22 | 6:38 | 70 | 83 | 72 | |
| 22 | Mon. | 5:23 | 6:37 | 74 | 80 | 70 | .80, very heavy rain 4 p.m. |
| 23 | Tues. | 5:24 | 6:36 | 78 | 86 | 71 | |
| 24 | Wed. | 5:25 | 6:35 | 70 | 88 | 77 | |
| 25 | Thurs. | 5:27 | 6:33 | 76 | 88 | 73 | .48, heavy shower 4 p.m. |
| 26 | Fri. | 5:28 | 6:32 | 70 | 86 | 78 | |
| 27 | Sat. | 5:29 | 6:31 | 72 | 85 | 76 | |
| 28 | Sun. | 5:30 | 6:30 | 72 | 78 | 71 | |
| 29 | Mon. | 5:31 | 6:29 | 65 | 80 | 68 | |
| 30 | Tues. | 5:32 | 6:28 | 60 | 80 | 68 | |
| 31 | Wed. | 5:33 | 6:27 | 62 | — | 64 | |

## September 1864

The Georgetown recorder's alignment of his second ledger page, recording precipitation and comments, continued from August through September. That makes those columns reliable for the dates recorded.

After "a great storm last night," a Richmond diarist wrote on September 6, "this morning it is very cool," and that tendency continued through the 9th. A "cloudy and threatening sky" on the 10th yielded "a heavy thunderstorm" on the 11th, during which "It rained hard." The next three days were "very cool" on the 12th, then "cold" on the morning of the 13th, and "very cool and cloudy" on the 14th. During the bitter fights at Fort Harrison and elsewhere east of Richmond at the end of the month, the diarist reported "beautiful weather but not as cool as in recent days" on the 28th. On the night of the 29th he mentioned "a very strong wind."[55]

"It rained without cessation during the night," an Alabama soldier marching with Early in the Shenandoah Valley wrote on September 10, and "we had a very damp time of it." Water dripped from the trees all night long onto his "head and face, and trickle[d] down my back." Another diarist with Early's army reported rain on September 4, 5, 6, 10, 12, 14, and 16.[56]

When the rain cleared off, troops had reason to appreciate the Valley as an improvement over Richmond's steamy environs. "The weather is cool and bracing," a general wrote. "It is, as you know, a pure, bracing climate." Ten days later, the same officer, having been ordered east across the mountains to Rapidan Station, rued the change: "The change from that side of the mountain to this is very marked. There, we were rather cold; here, decidedly summerish and lots of dust, and the pasturage is not as good and apples rather scarce."[57]

An officer in the lower Shenandoah Valley called Saturday the 17th "a glorious day." The same man, riding toward the Third Battle of Winchester, declared the 19th "a brilliant late September day."[58]

September 1864

| Date | Day of Week | Rich-mond Sunrise | Rich-mond Sunset | D.C. Temperature | | | Other |
|---|---|---|---|---|---|---|---|
| | | | | 7 a.m. | 2 p.m. | 9 p.m. | |
| 1 | Thurs. | 5:34 | 6:26 | 58 | 78 | 65 | |
| 2 | Fri. | 5:35 | 6:25 | 54 | 80 | 62 | |
| 3 | Sat. | 5:36 | 6:24 | 65 | — | 70 | |
| 4 | Sun. | 5:38 | 6:22 | 70 | 75 | 74 | slight shower 11 last night |
| 5 | Mon. | 5:39 | 6:21 | 71 | 78 | 63 | .36 |
| 6 | Tues. | 5:40 | 6:20 | 56 | 60 | 58 | drizzly from dark last night |
| 7 | Wed. | 5:41 | 6:19 | 54 | 70 | 63 | .80 yesterday & last night |
| 8 | Thurs. | 5:42 | 6:18 | 54 | 68 | 60 | drizzly all aft. & evening |
| 9 | Fri. | 5:44 | 6:16 | 60 | 75 | 78 | drizzly all night |
| 10 | Sat. | 5:45 | 6:15 | — | — | 62 | drizzle at 8 p.m. |
| 11 | Sun. | 5:46 | 6:14 | 64 | 78 | 60 | .46 |
| 12 | Mon. | 5:47 | 6:13 | 54 | 56 | 56 | shower thru the night |
| 13 | Tues. | 5:48 | 6:12 | 54 | 70 | 62 | |
| 14 | Wed. | 5:49 | 6:11 | 56 | 64 | 60 | .31, heavy shower at 3 |
| 15 | Thurs. | 5:50 | 6:10 | 60 | 78 | 70 | |
| 16 | Fri. | 5:51 | 6:09 | 58 | — | 60 | |
| 17 | Sat. | 5:53 | 6:07 | 50 | 78 | 67 | |
| 18 | Sun. | 5:54 | 6:06 | 64 | 78 | 66 | |
| 19 | Mon. | 5:55 | 6:05 | 62 | 74 | 58 | |
| 20 | Tues. | 5:56 | 6:04 | 50 | 78 | 64 | |
| 21 | Wed. | 5:58 | 6:02 | 58 | 80 | 70 | |
| 22 | Thurs. | 5:59 | 6:01 | 64 | 78 | 72 | slight shower last night |
| 23 | Fri. | 6:00 | 6:00 | 67 | 76 | 70 | Autumnal Equinox |
| 24 | Sat. | 6:01 | 5:59 | 72 | — | 69 | |
| 25 | Sun. | 6:03 | 5:57 | 54 | 66 | 56 | |
| 26 | Mon. | 6:04 | 5:56 | 52 | 76 | 66 | |
| 27 | Tues. | 6:05 | 5:55 | 58 | 78 | 68 | |
| 28 | Wed. | 6:07 | 5:53 | 64 | — | — | |
| 29 | Thurs. | 6:08 | 5:52 | 70 | 84 | 74 | |
| 30 | Fri. | 6:09 | 5:51 | 70 | 74 | 61 | drizzly at 7 |

## October 1864

October arrived in Richmond with cold rain. Overnight to the morning of October 1 "the weather was dreadful. It rained a lot." During the day "it was so cold" that a government clerk wore "my big cape for a cover" but still could not get warm. At 6 a.m. on the 4th he reported "thick fog," but the day turned into "rather good weather, hot." On October 7, General John Gregg died east of Richmond on the Darbytown Road on a day of "magnificent weather, bright sunlight and not too hot."[59]

A front brought in "cloudy, windy, and cold" weather to Richmond on October 8 and 9, and the first frost of the season appeared on the morning of the 10th. The rest of October featured many beautiful bright autumn days, with rain on the nights of the 12th and 21st.[60]

Near Ashby's Gap, a woman grumbled in her diary on October 6 of "a gloomy rainy week."[61]

In New Market on October 9, the weather was "cold, bitter cold. A little hail and a little snow fell yesterday and this morning."[62]

A member of General Early's staff called the historic day of battle at Cedar Creek, October 19, "a very fine day" that turned "cool at night." The 20th he described as "quite windy and cool" and the 21st as a "fine day, but cool."[63]

Southern survivors of the victory-turned-disaster at Cedar Creek, camped near New Market on October 23, found the "weather cold. . . . No blankets scarcely . . . the men having lost them in the late fight." Rain fell on New Market that night, and the 27th there was "cold and cloudy." A "perfect hurricane" roared through the Shenandoah Valley on the 28th, blowing campfires' "smoke and cinders in every direction." By the 30th good weather had returned, making "a warm, bright Sabbath . . . just such an autumn Sabbath as one loves to enjoy," a Vermont soldier wrote.[64]

Near Richmond, "a heavy storm" on the afternoon of October 27 brought with it heavy thunder and lightning. "Rain poured for seven hours" that night on a gunboat in the James River.[65]

## October 1864

| Date | Day of Week | Rich-mond Sunrise | Rich-mond Sunset | D.C. Temperature 7 a.m. | 2 p.m. | 9 p.m. | Other |
|------|------|------|------|------|------|------|------|
| 1 | Sat. | 6:10 | 5:50 | 54 | 56 | 58 | |
| 2 | Sun. | 6:11 | 5:49 | 61 | 71 | 62 | drizzle |
| 3 | Mon. | 6:12 | 5:48 | 59 | 70 | 66 | |
| 4 | Tues. | 6:13 | 5:47 | 60 | 74 | 68 | drizzly |
| 5 | Wed. | 6:14 | 5:46 | 68 | 76 | 69 | drizzled all day |
| 6 | Thurs. | 6:16 | 5:44 | 69 | — | 72 | |
| 7 | Fri. | 6:17 | 5:43 | 61 | 75 | 70 | |
| 8 | Sat. | 6:18 | 5:42 | 54 | 51 | 46 | 54 reading was at 4 a.m. |
| 9 | Sun. | 6:19 | 5:41 | 48 | 50 | 40 | |
| 10 | Mon. | 6:21 | 5:39 | 36 | 60 | 54 | first white frost |
| 11 | Tues. | 6:22 | 5:38 | 45 | 66 | 52 | |
| 12 | Wed. | 6:23 | 5:37 | 46 | 52 | 50 | 2:00 raining |
| 13 | Thurs. | 6:24 | 5:36 | 44 | 57 | 48 | |
| 14 | Fri. | 6:26 | 5:34 | 43 | 65 | 56 | |
| 15 | Sat. | 6:27 | 5:33 | 52 | 71 | 56 | |
| 16 | Sun. | 6:28 | 5:32 | 53 | 66 | 54 | |
| 17 | Mon. | 6:29 | 5:31 | 46 | — | 53 | |
| 18 | Tues. | 6:30 | 5:30 | 40 | 68 | 51 | |
| 19 | Wed. | 6:31 | 5:29 | 43 | — | 52 | |
| 20 | Thurs. | 6:32 | 5:28 | 42 | 68 | 44 | |
| 21 | Fri. | 6:33 | 5:27 | — | 62 | 50 | |
| 22 | Sat. | 6:34 | 5:26 | 42 | 53 | 44 | |
| 23 | Sun. | 6:36 | 5:24 | 48 | 59 | 46 | |
| 24 | Mon. | 6:37 | 5:23 | 40 | 62 | 53 | |
| 25 | Tues. | 6:38 | 5:22 | 42 | 64 | 46 | |
| 26 | Wed. | 6:39 | 5:21 | 36 | 63 | 52 | |
| 27 | Thurs. | 6:40 | 5:20 | 54 | 62 | 58 | |
| 28 | Fri. | 6:41 | 5:19 | 54 | 59 | 51 | |
| 29 | Sat. | 6:43 | 5:17 | 46 | 66 | 44 | |
| 30 | Sun. | 6:44 | 5:16 | 42 | — | 48 | |
| 31 | Mon. | 6:45 | 5:15 | 36 | 56 | 49 | white frost |

November 1864

Cold rain drenched Richmond all night on November 2–3 and again the next evening, which also turned windy. More rain fell during November 7–9.[66]

A soldier in camp near New Market did not enjoy November 2: "Horrible weather to-day. . . . It is *simply impossible* to keep warm." Rain and "fierce northwest winds prevail[ed]" on the 3rd; the sun shone on the 4th; snow fell on the 5th; and "the wind blew violently" on the night of the 6th, continuing through at least the 9th. Election day, November 8, came in foggy in the Shenandoah Valley despite the cold, and the day warmed markedly during the afternoon.[67]

In Richmond, November 15 dawned "very cold." By the 18th, "beautiful weather" had set in, with temperatures "a little milder." The next morning brought rain and "abominable weather . . . but it is not very cold." "Still raining" on the 21st, then not raining but "cold" on the 22nd. Snow overnight on November 22–23 did not amount to much, but "a very cold wind" blew on the 23rd. At 7:30 a.m. on the 26th, Richmond's temperatures were "cool, not very cold."[68]

Near Staunton, November 19 turned "rainy" with "some snow."[69]

Confederate cavalry and horse artillery fought a brisk skirmish on Rude's Hill and across Meem's Bottom in the Shenandoah Valley on November 22, "on a very cold day, when the ground was covered with snow, and a north wind blowing."[70]

## November 1864

| Date | Day of Week | Rich-mond Sunrise | Rich-mond Sunset | D.C. Temperature | | | Other |
|------|-------------|-------------------|------------------|---------|---------|---------|-------|
| | | | | 7 a.m. | 2 p.m. | 9 p.m. | |
| 1 | Tues. | 6:46 | 5:14 | 42 | 54 | 40 | |
| 2 | Wed. | 6:47 | 5:13 | 38 | 50 | 40 | drizzly from 4 p.m. |
| 3 | Thurs. | 6:48 | 5:12 | 39 | 46 | 44 | drizzled all day, all night |
| 4 | Fri. | 6:49 | 5:11 | 46 | 58 | 46 | cleared up at 8 a.m. |
| 5 | Sat. | 6:50 | 5:10 | 43 | 45 | 34 | |
| 6 | Sun. | 6:51 | 5:09 | 32 | 57 | 45 | heavy white frost |
| 7 | Mon. | 6:52 | 5:08 | 52 | 54 | 58 | rained all day and night |
| 8 | Tues. | 6:53 | 5:07 | — | 64 | 62 | drizzled all day |
| 9 | Wed. | 6:54 | 5:06 | 63 | 74 | 69 | |
| 10 | Thurs. | 6:55 | 5:05 | 68 | 66 | 48 | rained all last night |
| 11 | Fri. | 6:56 | 5:04 | 43 | 58 | 45 | |
| 12 | Sat. | 6:57 | 5:03 | 42 | — | 40 | |
| 13 | Sun. | 6:58 | 5:02 | 39 | 43 | 36 | |
| 14 | Mon. | 6:59 | 5:01 | 35 | 42 | 31 | |
| 15 | Tues. | 7:00 | 5:00 | 32 | — | 45 | |
| 16 | Wed. | 7:01 | 4:59 | 34 | 50 | 37 | |
| 17 | Thurs. | 7:01 | 4:59 | 41 | 52 | 50 | |
| 18 | Fri. | 7:02 | 4:58 | 53 | — | 47 | rained steadily all day |
| 19 | Sat. | 7:03 | 4:57 | 42 | — | 40 | |
| 20 | Sun. | 7:04 | 4:56 | 42 | 46 | 45 | drizzled all day |
| 21 | Mon. | 7:05 | 4:55 | 44 | 50 | 46 | rained all day |
| 22 | Tues. | 7:06 | 4:54 | — | 34 | 32 | |
| 23 | Wed. | 7:07 | 4:53 | 23 | 32 | 21 | |
| 24 | Thurs. | 7:07 | 4:53 | 24 | 40 | 30 | |
| 25 | Fri. | 7:08 | 4:52 | 24 | — | 35 | |
| 26 | Sat. | 7:09 | 4:51 | 34 | 48 | 45 | |
| 27 | Sun. | 7:09 | 4:51 | 38 | 56 | 38 | |
| 28 | Mon. | 7:10 | 4:50 | — | — | 50 | |
| 29 | Tues. | 7:11 | 4:49 | 50 | 70 | 56 | |
| 30 | Wed. | 7:11 | 4:49 | 51 | 70 | 50 | |

## December 1864

December 1 began "clear and warm" in Richmond, and three days later temperatures had turned only "a little cool" in the morning. December 6 proved again to be "a lovely day and still cool," but the 7th brought clouds and rain. Cold weather descended on the 8th, and through the 13th each day remained "extremely cold," "terribly cold," or "bitterly cold."[71]

Cold rain fell on Confederate camps near Harrisonburg on December 3, but good weather prevailed on the 5th. By the 10th, snow covered the ground there.[72]

A Virginian on the front lines near Richmond actually welcomed the advent of cold weather because of its impact on soldiers' health: "All of us were glad when cold weather came, because we hoped it would drive away the chills and fever of the summer, which to a certain extent it did."[73]

A Georgia soldier imprisoned at Point Lookout, Maryland, made diary entries about the weather there during December, where the adjacent Chesapeake Bay moderated temperatures a bit: 14th, "cloudy today. Rained some"; 16th, "cloudy today and warm"; 19th, "cloudy but no rain."; 21st, "In the night it rained, and most of today. . . . The water stands around our tent"; 22nd, "rained nearly all night. The blankets froze over us. Today the wind blows to our discomfort. The weather is about as cold as it ever gets to be."[74]

After constant rain the night of December 20–21, Richmond's weather turned into a pleasantly cold and clear Christmas week. Christmas Eve was "clear and cold," and Christmas Day "clear and pleasant—white frost." Rain began the night of the 25th and continued intermittently into the 29th.[75]

In recounting pursuit of a criminal into Gillie's Creek near Richmond on December 30, a local newspaper called the night, as might have been expected, "dark and cold," leaving both the captive and his captor "perfectly benumbed" by the excursion into a stream of running water.[76]

The popular illustrated newspapers brought Santa Claus into wide national circulation as a Christmas symbol during the Civil War. This scene also features holiday greenery and, in the background, snowy camp scenes, including snowballing soldiers. (*Harper's Weekly,* January 3, 1863, 1.)

## December 1864

| Date | Day of Week | Rich-mond Sunrise | Rich-mond Sunset | D.C. Temperature | | | Other |
|---|---|---|---|---|---|---|---|
| | | | | 7 a.m. | 2 p.m. | 9 p.m. | |
| 1 | Thurs. | 7:12 | 4:48 | 40 | 62 | 46 | |
| 2 | Fri. | 7:12 | 4:48 | 44 | 50 | 50 | |
| 3 | Sat. | 7:13 | 4:47 | 50 | 54 | 57 | |
| 4 | Sun. | 7:13 | 4:47 | 43 | 50 | 36 | |
| 5 | Mon. | 7:14 | 4:46 | 33 | 46 | 40 | |
| 6 | Tues. | 7:14 | 4:46 | 41 | 58 | 41 | |
| 7 | Wed. | 7:15 | 4:45 | 42 | 50 | 58 | 2:00 showery |
| 8 | Thurs. | 7:15 | 4:45 | 35 | 38 | 29 | high wind all night |
| 9 | Fri. | 7:15 | 4:45 | 19 | — | 26 | began to snow 8 p.m. |
| 10 | Sat. | 7:16 | 4:44 | 32 | — | — | snow 4 inches |
| 11 | Sun. | 7:16 | 4:44 | 34 | 38 | 27 | |
| 12 | Mon. | 7:16 | 4:44 | 14 | 24 | 19 | windy all night |
| 13 | Tues. | 7:17 | 4:43 | 20 | — | 33 | |
| 14 | Wed. | 7:17 | 4:43 | 38 | — | 42 | |
| 15 | Thurs. | 7:17 | 4:43 | 33 | 32 | 32 | snowing at 2 p.m. |
| 16 | Fri. | 7:17 | 4:43 | 32 | — | 34 | |
| 17 | Sat. | 7:18 | 4:42 | 34 | — | 37 | |
| 18 | Sun. | 7:18 | 4:42 | 39 | 44 | 40 | |
| 19 | Mon. | 7:18 | 4:42 | 40 | 51 | 44 | |
| 20 | Tues. | 7:18 | 4:42 | 30 | 36 | 28 | |
| 21 | Wed. | 7:18 | 4:42 | 30 | 34 | 34 | Winter Solstice |
| 22 | Thurs. | 7:18 | 4:42 | 27 | 26 | 20 | |
| 23 | Fri. | 7:18 | 4:42 | 19 | 30 | 24 | |
| 24 | Sat. | 7:18 | 4:42 | 28 | 31 | 31 | |
| 25 | Sun. | 7:18 | 4:42 | 21 | 44 | 42 | |
| 26 | Mon. | 7:18 | 4:42 | 38 | 43 | 40 | |
| 27 | Tues. | 7:17 | 4:43 | 37 | 44 | — | |
| 28 | Wed. | 7:17 | 4:43 | 43 | — | 44 | .65 rain yesterday & last night |
| 29 | Thurs. | 7:17 | 4:43 | 33 | 38 | 29 | windy all night |
| 30 | Fri. | 7:17 | 4:43 | 26 | — | 40 | snow last night 3 inches |
| 31 | Sat. | 7:17 | 4:43 | 34 | 30 | 25 | |

# Chapter 6

# 1865

## January 1865

The first day of 1865 came in "very cold" in Richmond, "and the ground is covered with snow. . . . Freezing cold weather." On January 4, "magnificent sunshine" reflected off a blanket of snow, making Richmond "very beautiful but not very pleasant." The next day more "lovely . . . bright sunshine" melted the snow and made "the roads . . . very slippery." The 7th and 8th both boasted "magnificent . . . beautiful sunlight. . . . Very cold but lovely."[1]

Troops mired in cold, damp trenches around Richmond accomplished little in any military sense through the winter. A newspaper reported an attempt to take advantage of bad weather: "the enemy, during the storm of [January 10], advanced their picket lines on our extreme right, south of Petersburg." The same storm "considerably damaged" some Confederate works near Petersburg but simultaneously interfered with Yankee attempts to dig the Dutch Gap canal below Richmond. On the 12th, the storm-engorged James River finally began to recede at about 4 p.m., subsiding more than fifteen inches in two hours.[2]

A Richmond diarist described January 12–16 as crisp and frosty, then clouds began "spitting snow" on the 17th. After another bright spell, January 21 proved to be "a dark, cold, sleety day, with rain." The next two days also brought "sleet and gloom" and fog and more rain.[3]

From his post in the trenches near Richmond, a captain in the 50th Georgia described January 15 in his diary as "a very pleasant day for this time of year." He wrote on the 21st of "a great deal of rain today, and it freezes as fast as it falls on the ground," and on the 23rd more rain fell "very nearly all day."[4]

Midwinter mud throttled military operations late in the month before they could begin. A correspondent with Lee's army opined on January 24 that "there is too much mud between the two armies for either to make a serious advance." Four days later he suggested that "the troops of the two armies have as much as they can do to keep from freezing."[5]

On January 29 a North Carolinian described the week just past on the front lines: "We had an awful time; the whole week it rained, and sleeted part of the time, and . . . it kept up the coldest wind that I ever felt. The men on vidette had to be relieved every half hour, to keep them from freezing. One man in our regiment got so cold he could hardly talk when relieved."[6]

## January 1865

| Date | Day of Week | Rich-mond Sunrise | Rich-mond Sunset | D.C. Temperature | | | Other |
|------|------|------|------|------|------|------|------|
| | | | | 7 a.m. | 2 p.m. | 9 p.m. | |
| 1 | Sun. | 7:16 | 4:44 | 22 | 29 | 19 | |
| 2 | Mon. | 7:16 | 4:44 | 18 | — | 28 | |
| 3 | Tues. | 7:15 | 4:45 | 22 | 35 | 29 | big snow at 3 p.m. |
| 4 | Wed. | 7:15 | 4:45 | 25 | 34 | 25 | snowed 3 inches |
| 5 | Thurs. | 7:15 | 4:45 | 14 | 36 | 36 | snowed |
| 6 | Fri. | 7:14 | 4:46 | 36 | 42 | 42 | rained all day |
| 7 | Sat. | 7:14 | 4:46 | 38 | 34 | 24 | .60 rain |
| 8 | Sun. | 7:13 | 4:47 | 16 | 27 | 23 | |
| 9 | Mon. | 7:12 | 4:48 | 25 | 36 | 36 | |
| 10 | Tues. | 7:12 | 4:48 | 38 | 42 | 50 | rained all last night |
| 11 | Wed. | 7:11 | 4:49 | 29 | 34 | 26 | high wind all night |
| 12 | Thurs. | 7:11 | 4:49 | 26 | 42 | 32 | windy night |
| 13 | Fri. | 7:10 | 4:50 | 36 | — | 40 | heavy white frost |
| 14 | Sat. | 7:09 | 4:51 | 40 | 37 | 32 | |
| 15 | Sun. | 7:09 | 4:51 | 32 | 40 | 33 | |
| 16 | Mon. | 7:08 | 4:52 | 28 | 32 | 26 | |
| 17 | Tues. | 7:07 | 4:53 | 24 | — | 28 | |
| 18 | Wed. | 7:06 | 4:54 | 22 | — | 22 | |
| 19 | Thurs. | 7:06 | 4:54 | 16 | 29 | 25 | |
| 20 | Fri. | 7:05 | 4:55 | 15 | — | 28 | heavy white frost |
| 21 | Sat. | 7:04 | 4:56 | 26 | 28 | 32 | hailing, rain, sleet, snow all day |
| 22 | Sun. | 7:03 | 4:57 | 31 | 34 | 33 | |
| 23 | Mon. | 7:03 | 4:57 | 34 | 36 | 34 | rain thru the day |
| 24 | Tues. | 7:02 | 4:58 | 27 | 34 | 23 | |
| 25 | Wed. | 7:01 | 4:59 | 15 | — | 18 | |
| 26 | Thurs. | 7:00 | 5:00 | 16 | 25 | 20 | |
| 27 | Fri. | 6:59 | 5:01 | 15 | 20 | 15 | |
| 28 | Sat. | 6:58 | 5:02 | 9 | 21 | 20 | quite windy at 4 p.m. |
| 29 | Sun. | 6:57 | 5:03 | 34 | 34 | 31 | |
| 30 | Mon. | 6:56 | 5:04 | 2- | 40 | 27 | |
| 31 | Tues. | 6:55 | 5:05 | 21 | 42 | 36 | |

## February 1865

The daring capture by Confederate raiders of Generals George Crook and Benjamin F. Kelley in Cumberland, Maryland, in mid-February became one of the most-discussed exploits of the war by irregular troops. The raiders' mountainous route took them through northern West Virginia. Since readings from the weather station in Cumberland itself obviously bear directly upon the conditions for the famous raid, this enumeration from that source covers the period (the station only recorded conditions once daily, at 7 a.m.): 20 degrees and fair on the 15th; 23 and fair on the 16th; 19 and dull on the 17th; 12 and fair on the 18th; 5 and fair on the 19th; 2 and fair on the 20th; 22 and rain (somehow not snow) on the 21st; 32 and dull on the 22nd; 35 and snow on the 23rd; 25 and windy on the 24th; and 8 and fair on the 25th. Confederates on the Cumberland raid obviously faced considerably colder temperatures than those during the more famous Romney campaign in January 1862 (*q.v.*).[7]

Brigadier General George T. "Tige" Anderson's diary recorded cold, rain, and sleet on February 3, 5, and 7. The bad weather affected the fighting at Hatcher's Run on February 6 and 7, which killed newlywed General John Pegram. A correspondent with the army wrote on the 7th: "the severe snow and sleet . . . now bids fair to put an end to military operations on these lines."[8]

A Virginian defending the Howlett Line southeast of Richmond remembered this winter's weather as "intensely cold, the ground covered with ice and snow." Needing wood for fires and warmth, when everything behind the lines had been cleared for a mile, the pickets from both sides shared "some good wood" between the lines, "by mutual agreement and concession."[9]

Winter's frigid grip on central Virginia turned Richmond's streets into glistening rinks. Newspapers reported serious injury to a woman who fell on an icy street at the market, and on February 14 a convalescing soldier fell on ice on Main Street, reopening his wound.[10]

February 15 dawned "sleety and dangerous" in Richmond, then turned rainy through the next night. After more rain on February 16 and 17, the 18th proved to be "as lovely a morning as ever dawned on earth. A gentle southern

Wintry sentry duty. (Alfred H. Guernsey and Henry M. Alden, eds., *Harper's Pictorial History of the Civil War* [Chicago: Star Publishing Company, 1894], 524.)

breeze, a cloudless sky, and a glorious morning sun . . . dispels the moisture of the late showers."[11]

When the cold abated, thawing ground and "drenching rain" on February 23 "converted Eastern Virginia into one vast quagmire." More rain on the 25th thwarted a reported Yankee advance.[12]

## February 1865

| Date | Day of Week | Rich-mond Sunrise | Rich-mond Sunset | D.C. Temperature | | | Other |
|------|-------------|-------------------|------------------|--------|--------|--------|-------|
| | | | | 7 a.m. | 2 p.m. | 9 p.m. | |
| 1 | Wed. | 6:54 | 5:06 | 34 | — | 34 | |
| 2 | Thurs. | 6:53 | 5:07 | 30 | — | 32 | |
| 3 | Fri. | 6:52 | 5:08 | 25 | — | 32 | heavy white frost |
| 4 | Sat. | 6:51 | 5:09 | 34 | 50 | 38 | smoky & foggy |
| 5 | Sun. | 6:50 | 5:10 | 40 | 38 | 27 | |
| 6 | Mon. | 6:49 | 5:11 | 23 | 34 | 27 | |
| 7 | Tues. | 6:48 | 5:12 | 27 | 26 | 32 | snowed 6 inches |
| 8 | Wed. | 6:47 | 5:13 | 26 | — | 26 | |
| 9 | Thurs. | 6:46 | 5:14 | 25 | 30 | 22 | |
| 10 | Fri. | 6:45 | 5:15 | — | 40 | 34 | |
| 11 | Sat. | 6:44 | 5:16 | 29 | 38 | 28 | |
| 12 | Sun. | 6:43 | 5:17 | 28 | 28 | 18 | snowed .5 inch last night |
| 13 | Mon. | 6:42 | 5:18 | 6 | 22 | 12 | |
| 14 | Tues. | 6:41 | 5:19 | 6 | 3- | 27 | |
| 15 | Wed. | 6:40 | 5:20 | 29 | 34 | 35 | rained & sleeted all day & windy |
| 16 | Thurs. | 6:39 | 5:21 | 33 | 37 | 35 | |
| 17 | Fri. | 6:37 | 5:23 | 28 | 40 | 36 | heavy fog, rain |
| 18 | Sat. | 6:36 | 5:24 | 34 | 44 | 38 | |
| 19 | Sun. | 6:35 | 5:25 | 36 | 44 | 33 | |
| 20 | Mon. | 6:34 | 5:26 | 28 | — | 31 | |
| 21 | Tues. | 6:32 | 5:28 | 24 | 44 | 33 | heavy white frost |
| 22 | Wed. | 6:31 | 5:29 | 29 | 44 | 42 | heavy white frost |
| 23 | Thurs. | 6:30 | 5:30 | 44 | 48 | 43 | thick fog all p.m. |
| 24 | Fri. | 6:29 | 5:31 | 34 | 49 | 37 | |
| 25 | Sat. | 6:28 | 5:32 | 33 | 36 | 36 | rained all day from 10 a.m. |
| 26 | Sun. | 6:27 | 5:33 | 39 | 60 | 50 | |
| 27 | Mon. | 6:26 | 5:34 | 41 | 59 | 42 | |
| 28 | Tues. | 6:25 | 5:35 | 38 | 56 | 37 | |

## March 1865

General Jubal A. Early's tiny fragment fought the final action in the Shenandoah Valley on March 2 at Waynesboro. March 1 had been "pleasant but chilly" there, but "cold sleet . . . constantly falling" pelted the region on the 2nd. "The trees were loaded down with sleet and the ground covered with snow."[13]

Late February's mud carried across into a sticky early March in the armies' lines around the Confederate capital. On March 2, 3, and 4 the Richmond newspapers reported that mud ruled the battlefront. "The mud keeps everything at a standstill in the armies before Richmond and Petersburg." "The rain and mud lasts, and of these there seems to be no end." March 4–7 remained dry in Richmond, but "heavy rain" on the 8th muddied up the countryside anew. "Fine weather" prevailed on Tuesday the 14th.[14]

A Confederate bureau official wrote gloomily early in this month, "The month of March is a vile month in Richmond."[15]

Richmond's residents saw a "splendid rainbow" on the night of March 15. The next night "a violent southeast gale prevailed . . . with rain." Bright sun tempered winds still blowing through Richmond on the 18th, but dawn of the 19th reminded a diarist of spring in the Garden of Eden. Warm and pleasant weather continued into the 21st, when blossoms appeared on apricot trees.[16]

A staff officer engaged in the assault on Fort Stedman described the early morning of March 25 as "grey, cold and gloomy."[17]

Heavy rains late in March filled the James River with so much water that the "islands were inundated" late on the 31st. The river level "bid[s] fair to exceed the highest we have had for years," a local man wrote.[18] A few hours later, Confederates would be scampering across the few available bridges spanning the flooded river as they evacuated Richmond.

## March 1865

| Date | Day of Week | Rich-mond Sunrise | Rich-mond Sunset | D.C. Temperature | | | Other |
|------|------|------|------|------|------|------|------|
| | | | | 7 a.m. | 2 p.m. | 9 p.m. | |
| 1 | Wed. | 6:23 | 5:37 | 34 | 44 | 38 | |
| 2 | Thurs. | 6:22 | 5:38 | 34 | 40 | 40 | |
| 3 | Fri. | 6:21 | 5:39 | 39 | 44 | 44 | drizzled all day |
| 4 | Sat. | 6:20 | 5:40 | 42 | 64 | 40 | rain all night, cleared 10 a.m. |
| 5 | Sun. | 6:19 | 5:41 | 32 | 44 | 33 | |
| 6 | Mon. | 6:17 | 5:43 | 28 | 49 | 36 | white frost |
| 7 | Tues. | 6:16 | 5:44 | 32 | — | 43 | heavy white frost |
| 8 | Wed. | 6:15 | 5:45 | 39 | 62 | 54 | began to rain 5 p.m. |
| 9 | Thurs. | 6:14 | 5:46 | 54 | 67 | 58 | rained all night |
| 10 | Fri. | 6:12 | 5:48 | 40 | 34 | 28 | |
| 11 | Sat. | 6:11 | 5:49 | 22 | 43 | 34 | |
| 12 | Sun. | 6:10 | 5:50 | 39 | 59 | 48 | |
| 13 | Mon. | 6:09 | 5:51 | 56 | 68 | 56 | |
| 14 | Tues. | 6:08 | 5:52 | 48 | 72 | 58 | wild geese passed north |
| 15 | Wed. | 6:07 | 5:53 | 59 | 66 | 61 | rainlike |
| 16 | Thurs. | 6:06 | 5:54 | 62 | 72 | 68 | drizzle at 7, flying clouds & occ. sun |
| 17 | Fri. | 6:04 | 5:56 | 38 | 59 | 54 | heavy rain & wind all night |
| 18 | Sat. | 6:03 | 5:57 | 51 | 57 | 44 | wind blustery all day |
| 19 | Sun. | 6:02 | 5:58 | 46 | 60 | 49 | |
| 20 | Mon. | 6:00 | 6:00 | 44 | 72 | 60 | Vernal Equinox |
| 21 | Tues. | 5:59 | 6:01 | 56 | 75 | 51 | began to rain 6 p.m. |
| 22 | Wed. | 5:58 | 6:02 | 51 | 55 | 50 | |
| 23 | Thurs. | 5:57 | 6:03 | 46 | 55 | 44 | 2 p.m. thunder, high wind |
| 24 | Fri. | 5:55 | 6:05 | 36 | 47 | 42 | windy all day |
| 25 | Sat. | 5:54 | 6:06 | 42 | 47 | 41 | sprinkled all day |
| 26 | Sun. | 5:53 | 6:07 | 36 | 46 | 33 | apricots in bloom |
| 27 | Mon. | 5:52 | 6:08 | 36 | 57 | 45 | |
| 28 | Tues. | 5:51 | 6:09 | 42 | 62 | 50 | |
| 29 | Wed. | 5:50 | 6:10 | 46 | 72 | 61 | sprinkly occasionally |
| 30 | Thurs. | 5:49 | 6:11 | 56 | 66 | 56 | drizzle |
| 31 | Fri. | 5:48 | 6:12 | 50 | 46 | 50 | |

April 1865

Richmond's final moments as capital of the Confederate States unfolded under pleasant, spring-like weather conditions. "It is beautiful weather," a Richmond diarist wrote on the morning of April 1, and the next day he enjoyed sitting in the sun under cool temperatures. A "strong wind" spurred the disastrous Richmond fire of April 3. On the momentous morning of April 9, Richmond was "a little cool." After a rainy evening on the 10th, clouds covered the sky on the morning of the 11th. Heavy rain returned the night of April 12 and continued the next day. After a brightly sunlit April 14, heavy rain fell on the 15th.[19]

Outside Richmond, where Lee's lines soon would be broken, an artillerist wrote in his diary on the morning of Sunday, April 2: "The weather is fine and there is not a cloud in the sky." When he arose on the 4th, along the army's route of retreat, he wrote: "We are having a rainy day. The roads are right muddy." Seven days later, as a surrendered ex-Confederate near Appomattox, the same diarist wrote: "there is much mud all around us."[20]

On April 10 in Nelson County, a homeward-bound Confederate recalled, "early in the day it set to raining. We traveled all day through a downpour." A Confederate general awaiting parole at Appomattox described the weather on April 10 and 11 in his diary: "It rained, and . . . the men were exposed to a pitiless storm, without shelter."[21]

A newly surrendered, if not yet reconstructed, rebel wrote in his diary near City Point on April 13, "the nights are dismally cold and cloudy . . . it having rained nearly every day since our capture." Through the night of the 12th, "the rain fairly poured down . . . *a perfect deluge.* . . . Oh, thou black, horrible, cruel night."[22]

Confederates imprisoned by the thousands at Point Lookout, Maryland, benefited from the advent of warmer weather. In a letter dated April 27, a Virginian reveled, despite his captive status, in the "balmy zephyrs and the gentle ripple of the waves of the beautiful Chesapeake Bay."[23]

April 1865

| Date | Day of Week | Rich-mond Sunrise | Rich-mond Sunset | D.C. Temperature | | | Other |
|------|-------------|-------------------|------------------|-------|--------|--------|-------|
| | | | | 7 a.m. | 2 p.m. | 9 p.m. | |
| 1 | Sat. | 5:46 | 6:14 | 49 | 61 | 45 | |
| 2 | Sun. | 5:45 | 6:15 | 47 | 63 | 56 | |
| 3 | Mon. | 5:44 | 6:16 | 49 | 62 | — | |
| 4 | Tues. | 5:43 | 6:17 | 48 | 63 | 54 | |
| 5 | Wed. | 5:41 | 6:19 | 53 | 70 | 60 | |
| 6 | Thurs. | 5:40 | 6:20 | 59 | 72 | 66 | cherry, plum, peach in blossom |
| 7 | Fri. | 5:39 | 6:21 | 64 | 69 | 51 | occasional rain all day |
| 8 | Sat. | 5:38 | 6:22 | 44 | 68 | 47 | |
| 9 | Sun. | 5:36 | 6:24 | 46 | 57 | 50 | |
| 10 | Mon. | 5:35 | 6:25 | 44 | 50 | 48 | rain thru night, drizzly all day |
| 11 | Tues. | 5:34 | 6:26 | 50 | 58 | — | |
| 12 | Wed. | 5:33 | 6:27 | 60 | 78 | 62 | showery all day, heavy 6 p.m. |
| 13 | Thurs. | 5:32 | 6:28 | 51 | 68 | 52 | |
| 14 | Fri. | 5:31 | 6:29 | 47 | 67 | 54 | |
| 15 | Sat. | 5:30 | 6:30 | 50 | 57 | 58 | drizzly |
| 16 | Sun. | 5:29 | 6:31 | 60 | 68 | 47 | Easter Sunday |
| 17 | Mon. | 5:28 | 6:32 | 41 | 57 | 49 | |
| 18 | Tues. | 5:26 | 6:34 | 52 | — | 61 | |
| 19 | Wed. | 5:25 | 6:35 | 58 | — | 59 | |
| 20 | Thurs. | 5:24 | 6:36 | 54 | 58 | 54 | |
| 21 | Fri. | 5:23 | 6:37 | 52 | 65 | 61 | |
| 22 | Sat. | 5:22 | 6:38 | 67 | 70 | 59 | |
| 23 | Sun. | 5:21 | 6:39 | 48 | 53 | 43 | |
| 24 | Mon. | 5:20 | 6:40 | 48 | 68 | 50 | |
| 25 | Tues. | 5:19 | 6:41 | 50 | 80 | 60 | |
| 26 | Wed. | 5:18 | 6:42 | 54 | 85 | 65 | |
| 27 | Thurs. | 5:17 | 6:43 | 61 | 86 | 68 | |
| 28 | Fri. | 5:16 | 6:44 | 64 | — | 66 | |
| 29 | Sat. | 5:15 | 6:45 | — | 80 | 56 | windy thru p.m., rained 4–8 |
| 30 | Sun. | 5:14 | 6:46 | 64 | 66 | 54 | murky sky, rainy at dark |

May 1865

May began with a "quite cool" morning in Richmond, after rain all night long. May 2 was "very cold," and the next day continued "quite cool"; but the 3rd was not as cool, "and is beautiful weather." Rain all night on the 11th turned into a "very beautiful and fresh" May 12.[24]

An ex-Confederate Virginian languishing as a prisoner near Newport News described May 5 in his diary as "a gloomy, rainy day." The 7th proved to be lovely, but on the 8th "a severe thunder-storm" blew down tents and scattered everything. "Chilly winds" swept in on the 12th, but warm sunshine returned the next day. Cloudy weather on the 20th turned into rain that began on the 21st and continued into that night, a storm that "raged . . . with great fury and violence. . . . I never saw such flashes of lightning before." He concluded that what Virginians call "the long spell in May is upon us." Lovely weather returned on the 23rd, but the next day turned "very cold . . . for the season." More rain fell on the night of May 25 and on into the 27th.[25]

Several mid-month comments on the second page of the Georgetown ledger fall far enough out of alignment to be of uncertain date; they are not reported at all in the May table.

May 1865

| Date | Day of Week | Rich-mond Sunrise | Rich-mond Sunset | D.C. Temperature | | | Other |
|---|---|---|---|---|---|---|---|
| | | | | 7 a.m. | 2 p.m. | 9 p.m. | |
| 1 | Mon. | 5:13 | 6:47 | 50 | 58 | 47 | |
| 2 | Tues. | 5:12 | 6:48 | 48 | 64 | 49 | |
| 3 | Wed. | 5:11 | 6:49 | 53 | 67 | 54 | loud thunder, rain last night |
| 4 | Thurs. | 5:10 | 6:50 | 60 | 70 | 60 | |
| 5 | Fri. | 5:09 | 6:51 | 76 | 77 | 62 | |
| 6 | Sat. | 5:08 | 6:52 | 66 | 80 | 76 | |
| 7 | Sun. | 5:07 | 6:53 | 64 | 71 | 62 | |
| 8 | Mon. | 5:06 | 6:54 | 60 | 82 | 68 | showery all day |
| 9 | Tues. | 5:05 | 6:55 | 70 | 62 | 66 | rained all night |
| 10 | Wed. | 5:04 | 6:56 | 55 | 64 | 58 | |
| 11 | Thurs. | 5:03 | 6:57 | 62 | 83 | 60 | |
| 12 | Fri. | 5:02 | 6:58 | 50 | 63 | 50 | |
| 13 | Sat. | 5:01 | 6:59 | 49 | 70 | 54 | |
| 14 | Sun. | 5:00 | 7:00 | 57 | 74 | 62 | |
| 15 | Mon. | 4:59 | 7:01 | 54 | 77 | 62 | |
| 16 | Tues. | 4:58 | 7:02 | 60 | 79 | 69 | |
| 17 | Wed. | 4:57 | 7:03 | 68 | 86 | 72 | |
| 18 | Thurs. | 4:57 | 7:03 | 74 | 84 | 62 | |
| 19 | Fri. | 4:56 | 7:04 | 57 | 70 | 67 | |
| 20 | Sat. | 4:55 | 7:05 | 69 | 78 | — | |
| 21 | Sun. | 4:55 | 7:05 | 6- | 7- | 72 | |
| 22 | Mon. | 4:54 | 7:06 | 65 | 84 | 72 | |
| 23 | Tues. | 4:53 | 7:07 | 68 | 75 | 62 | |
| 24 | Wed. | 4:53 | 7:07 | 68 | 75 | 62 | |
| 25 | Thurs. | 4:52 | 7:08 | 62 | 75 | 67 | |
| 26 | Fri. | 4:51 | 7:09 | 60 | 66 | 56 | rained all day |
| 27 | Sat. | 4:51 | 7:09 | 52 | 64 | 58 | drizzly all day |
| 28 | Sun. | 4:50 | 7:10 | 62 | 69 | 60 | |
| 29 | Mon. | 4:49 | 7:11 | 67 | 78 | 64 | |
| 30 | Tues. | 4:49 | 7:11 | 66 | 84 | 69 | |
| 31 | Wed. | 4:48 | 7:12 | 69 | 88 | 68 | |

June 1865

In Fairfax County, June opened warm. "Summer has come upon us all at once," a woman living near Mount Vernon wrote on June 2.[26]

Confederate prisoners being held near Norfolk wilted under a "hot, broiling sun" as they stood for long hours at a mandatory muster on June 4. The men without shoes attempted to protect their tender feet from sunburn by covering them with grass. The field expedient did not work for a Virginia diarist: "The top of my feet is covered with a solid scab—the effect of the sun." Blessed cool weather arrived on the 8th, "a delightful season." Four days later "the scorching sun" had returned to "shining 'cussedly' hot."[27]

Travelers going down the James River from Richmond to Yorktown on June 17 complained of the oppressive heat: "very hot. . . . very hot at the height of the day," but then "the evening was quite cool." Off Fort Monroe on the 18th, the sun rose on "a beautiful day." On the 22nd, passengers on board a vessel in Norfolk harbor watched "a very beautiful sunrise"; after dark, the decks felt chilly.[28]

Around Ashby's Gap, "rain pour[ed] all morning" on June 20, then cleared off at noon. That evening "the wind blew a perfect gale."[29]

June 1865

| Date | Day of Week | Rich-mond Sunrise | Rich-mond Sunset | D.C. Temperature | | | Other |
|------|------|------|------|------|------|------|------|
| | | | | 7 a.m. | 2 p.m. | 9 p.m. | |
| 1 | Thurs. | 4:48 | 7:12 | 68 | 86 | 78 | |
| 2 | Fri. | 4:47 | 7:13 | 71 | 89 | 72 | |
| 3 | Sat. | 4:47 | 7:13 | 70 | 8- | 74 | |
| 4 | Sun. | 4:46 | 7:14 | 80 | 90 | 80 | |
| 5 | Mon. | 4:46 | 7:14 | 75 | 92 | 82 | |
| 6 | Tues. | 4:45 | 7:15 | 65 | 71 | 66 | |
| 7 | Wed. | 4:45 | 7:15 | 65 | 86 | 73 | |
| 8 | Thurs. | 4:44 | 7:16 | 75 | 88 | 76 | |
| 9 | Fri. | 4:44 | 7:16 | 76 | 90 | 79 | |
| 10 | Sat. | 4:44 | 7:16 | 76 | 74 | 69 | |
| 11 | Sun. | 4:44 | 7:16 | 68 | 80 | 67 | |
| 12 | Mon. | 4:43 | 7:17 | 70 | 84 | 76 | |
| 13 | Tues. | 4:43 | 7:17 | 77 | 87 | 70 | heavy shower 5:30 p.m. |
| 14 | Wed. | 4:43 | 7:17 | 70 | 82 | 70 | heavy shower 4 p.m. |
| 15 | Thurs. | 4:43 | 7:17 | 67 | — | 68 | |
| 16 | Fri. | 4:42 | 7:18 | 68 | 82 | 73 | |
| 17 | Sat. | 4:42 | 7:18 | — | 88 | — | |
| 18 | Sun. | 4:42 | 7:18 | 85 | 87 | 75 | |
| 19 | Mon. | 4:42 | 7:18 | 78 | 90 | 72 | heavy shower |
| 20 | Tues. | 4:42 | 7:18 | 76 | 72 | — | thunder, lightning, heavy rain at 2 |
| 21 | Wed. | 4:42 | 7:18 | 70 | 85 | 72 | Summer Solstice |
| 22 | Thurs. | 4:42 | 7:18 | 72 | 83 | 74 | |
| 23 | Fri. | 4:42 | 7:18 | 76 | 90 | 74 | |
| 24 | Sat. | 4:42 | 7:18 | 75 | 8- | 78 | |
| 25 | Sun. | 4:42 | 7:18 | 78 | 87 | 76 | |
| 26 | Mon. | 4:42 | 7:18 | 73 | 80 | 66 | |
| 27 | Tues. | 4:43 | 7:17 | 67 | 82 | 72 | |
| 28 | Wed. | 4:43 | 7:17 | 72 | 84 | 74 | |
| 29 | Thurs. | 4:43 | 7:17 | 80 | 91 | 80 | |
| 30 | Fri. | 4:43 | 7:17 | 76 | 94 | 71 | heavy shower at 5 |

# Notes

## Introduction

1. Douglas Southall Freeman, "An Address," *Civil War History* 1 (1955): 10–11. Freeman delivered his remarks to a tour group from the Chicago Civil War Round Table on May 7, 1953. They appeared as the very first item in the first issue of *Civil War History*.

2. The Weather Bureau's typed synopsis of Mackee's ledgers, at the head of the filmed copy, mistakenly locates the readings at Lewinsville through October 1860, and the first month at Georgetown as November 1860. The original documents, however, clearly show the shift to Georgetown in November 1859. Furthermore, Mackee was enumerated in Georgetown by the 1860 census, which was taken five months before November 1860. Although the scope of this book extends only to June 1865, Mackee's ledgers continue to May 1866. The 1860 census and the Weather Bureau cover material are cited below in notes 4 and 5, respectively.

3. In his weather notes he wrote his name repeatedly as "Mackee," with a lower-case "k."

4. The Mackee household appears on page 106 of Georgetown, Ward Three, of the Eighth Census of the United States, M653, Roll 101, National Archives, Washington, D.C.

5. The title page at the front of the microfilm roll that contains Mackee's ledgers bears the heading "U.S. Department of Commerce, Weather Bureau, National Weather Records Center, Federal Building, Asheville, N.C. 28801." There is no hint of either Smithsonian or Signal Corps involvement in Mackee's manuscripts, although some of the other stations cited elsewhere in this book use forms that bear indications of those organizations. Although there is no date on the film that includes Georgetown's observations, most of the other stations used in this book were copied in 1952, for instance, the Richmond manuscripts (see the narrative covering October 1860).

6. The full title of Darter's work, which runs to 222 pages (including lxii as roman numerals), is *The National Archives: List of Climatological Records in the National Archives* (Washington, D.C.: National Archives, 1942). The verso of the title page reports the publication as "The National Archives Special List No. 1." The tables in this book use reports from two of those three Virginia stations (Alexandria and Richmond) to cover the brief Georgetown gaps.

7. Darter, *List of Climatological Records,* xlvi–xlix. The National Archives internet site (www.archives.gov) offers a current finding aid to Record Group 27. It obviously affords a more current overview than that in Darter's book, but without nearly as much detail (it covers only twelve pages). The modern designation for the Smithsonian records within the record groups is Record Group 27.3. In April 2006 a Smithsonian archivist reiterated to me the thorough transfer of weather material to the National Archives, and cited only "a small amount of material" surviving in their own Record Unit 60.

8. For the full text of this insightful comment (by a Richmond Howitzers soldier) about weather and diaries, and citation to the original, see the seventh entry for April 1862 and its corresponding note.

## Chapter 1. 1860

The times of sunrise and sunset in the tables for 1860 come from *(Cottom's Edition) Richardson's Almanac, 1860 . . . Calculated by David Richardson, of Louisa County, Va.* (Richmond: J. W. Randolph, [1859]).

1. Elizabeth Lindsay Lomax, *Leaves from an Old Washington Diary* (Mount Vernon, N.Y.: S. A. Jacobs, Golden Eagle Press, 1943), 130–32.

2. The manuscript Richmond ledgers were filmed in 1952 for the weather archives in Asheville under the heading "The United States Weather Bureau, Climatological Records, 1819–1892, National Archives Disposal Job No. III-NNR-12." A few gaps appear in some months of the Richmond run (through February 1862), and all of June, August, and December 1861 and January 1862 are missing. The 1860 census (Henrico County, Western Division, page 916) reported Meriwether as age sixty-two, born in Virginia, married to Louisa M. (age fifty-seven), and owning $20,000 in real estate and $14,500 in personal estate.

3. A handful of weather ledgers from Fredericksburg survive from this period. They were copied in the same 1952 National Archives project that made the Richmond film cited above. The "violent storm" quote and 4-to-midnight timing are from that manuscript. The "cruel storm" quote is from the *Fredericksburg News,* October 19, 1860—an evening paper, reporting on very recent events, and timing the onset as 1 p.m.

4. *Fredericksburg News,* November 2, 1860.

5. Betsy Fleet and John D. P. Fuller, eds., *Green Mount: A Virginia Plantation*

*Family during the Civil War, Being the Journal of Benjamin Robert Fleet and Letters of His Family* (Lexington: University of Kentucky Press, 1962), 39, 41.

6. The Smithfield weather manuscripts cover only a few months ending in March 1861. They were collected in the same 1952 National Archives job for the Asheville weather archives cited in note 2 above for Richmond.

7. Lomax, *Washington Diary*, 133.

8. Ada W. Bacot, *A Confederate Nurse* (Columbia: University of South Carolina Press, 1994), 18–19.

9. For citation of the Richmond weather ledgers, see note 2 above.

10. Lomax, *Washington Diary*, 134–36.

11. *Fredericksburg News,* December 4, 1860.

12. Fleet and Fuller, *Green Mount,* 42–43.

13. Bacot, *A Confederate Nurse,* 20.

14. Note 2 above supplies the citation for the Richmond weather source.

15. For the manuscript Fredericksburg weather source, see the citation in note 3 above.

16. Meticulously detailed weather readings made at the Virginia Military Institute (VMI) in Lexington survive for January, February, and March 1861. They were copied in the same 1952 National Archives job that preserved so many other ledgers of this sort (see note 2 above). The late-December blizzard is noted at the head of the January 1861 sheet from VMI; the December sheet does not survive, at least not in this grouping.

## Chapter 2. 1861

The times of sunrise and sunset in the tables for 1861 come from *(Cottom's Edition) Richardson's Almanac, 1861 . . . Calculated by David Richardson, of Louisa County, Va.* (Richmond: J. W. Randolph, [1860]).

1. Lomax, *Washington Diary*, 138–41.

2. For citation of the Richmond weather sheets, see chapter 1, note 2.

3. *Richmond Dispatch,* January 12, 1861.

4. See chapter 1, note 2 for December 1860 for the source of Lexington weather manuscripts.

5. *Fredericksburg News,* January 14, 1861.

6. *Richmond Dispatch,* January 21, 1861.

7. Lomax, *Washington Diary*, 141–44.

8. See chapter 1, note 2 for the source of the Richmond weather manuscripts.

9. Amanda V. Edmonds, *Journals of Amanda Virginia Edmonds, Lass of the Mosby Confederacy, 1859–1867,* ed. Nancy Chappelear Baird (Stephens City, Va.: Commercial Press, 1984), 46.

10. These February-long Richmond summaries include entries from February 1–6

and February 28, days not shown on this month's table because the Georgetown station operated during those seven dates.

11. The scant data from Smithfield are offered here not only for their own modest immediate interest but also for comparison to Richmond and Washington. The relationship for this month and the next (the only full months available for Smithfield) conforms to predictable expectations of difference. See chapter 1, note 6 for citation of the Smithfield source.

12. *Richmond Dispatch,* February 7, 1861.

13. See chapter 1, note 16 for the source of Lexington weather reports.

14. *Fredericksburg News,* March 1, 1861.

15. Lomax, *Washington Diary,* 144–46.

16. Chapter 1, note 2 supplies the citation for Richmond's weather report.

17. See chapter 1, note 16 for citation of the source of Lexington weather readings.

18. For an earlier aurora borealis in Fredericksburg, and for citations to that city's weather records, from which these March reports come, see chapter 1, notes 3 and 15.

19. The Smithfield weather source is cited in chapter 1, note 6.

20. These April figures all come from the same Richmond manuscripts cited in chapter 1, note 2.

21. Edmonds, *Journals,* 46.

22. *Fredericksburg News,* April 9, 12, 1861; *Richmond Dispatch,* April 15, 1861.

23. Lomax, *Washington Diary,* 147–51.

24. *Richmond Dispatch,* April 16, 18, 1861.

25. *Richmond Dispatch,* April 24, 27, 1861.

26. William S. White diary entry in Carlton McCarthy, ed., *Contributions to a History of the Richmond Howitzer Battalion,* 4 vols. (Richmond: Carlton McCarthy & Co., 1883–86), 2:93.

27. *Richmond Dispatch,* May 4, 1861.

28. Joseph A. Graves, *The History of the Bedford Light Artillery* (Bedford City, Va.: Press of the Bedford Democrat, 1903), 11.

29. W. T. Price, *"On to Grafton"* (Marlinton, W.Va.: privately printed, 1901), 36, 38. All quotations from this work come from the pages that reproduce the contemporary diaries of Osborne Wilson and Charles Lewis Campbell.

30. Emma C. R. Macon and Reuben C. Macon, *Reminiscences of the Civil War* (Cedar Rapids, Iowa: Torch Press, 1911), 143.

31. *Richmond Dispatch,* June 4, 1861.

32. For the source of Richmond weather reports, see chapter 1, note 2.

33. John B. Jones, *A Rebel War Clerk's Diary at the Confederate States Capitol,* 2 vols. (Philadelphia: J. B. Lippincott & Co., 1866), 1:34.

34. Price, *"On to Grafton,"* 40, 42.

35. *Richmond Dispatch,* June 10, 1861.

36. White diary in McCarthy, *Richmond Howitzer Battalion,* 2:95, 98.

37. *Richmond Dispatch,* June 25, 1861. The chances that a dozen miles affected the temperature are negligible; soldiers' fates always seem worse than those of their mates.

38. [John H. Grabill], *Diary of a Soldier in the Stonewall Brigade* (Woodstock, Va.: Press of Shenandoah Herald, 1909), entries (printed on unnumbered leaves) for June 30–July 2.

39. Kittrell J. Warren, *History of the Eleventh Georgia Vols.* (Richmond: Smith, Bailey & Co., 1863), 28.

40. *A Narrative of the Battles of Bull Run and Manassas Junction, July 18th and 21st, 1861* (Charleston, S.C.: Steam-Power Presses of Evans & Cogswell, 1861), 10.

41. George T. Todd, *Sketch of History, The First Texas Regiment* (n.p, n.d.), second and third unnumbered leaves.

42. Cadmus M. Wilcox manuscript memoir, University of Texas, Austin.

43. Douglas Lee Gibboney, ed., *Littleton Washington's Journal* (n.p.: Xlibris Corporation, 2001), 195; James McHenry Howard, *Recollections* (Baltimore: The Sun Book and Job Printing Office Inc., 1922), 41. The Howard book should not be confused with the much more familiar memoir by his relative McHenry Howard.

44. Price, *"On to Grafton,"* 54.

45. Edmonds, *Journals,* 55.

46. All Richmond readings are from the manuscript cited in chapter 1, note 2.

47. Warren, *Eleventh Georgia,* 29. Anderson's mid-month weather complaints cover his diary spaces for August 16–19; copy of the diary in the author's possession.

48. Laura Elizabeth Lee, ed., *Forget-Me-Nots of the Civil War* (St. Louis: Press A. R. Fleming Printing Co., 1909), 48.

49. Price, *"On to Grafton,"* 55.

50. Robert E. L. Strider, *The Life and Work of George William Peterkin* (Philadelphia: George W. Jacobs & Company, 1929), 47; John H. Worsham, *One of Jackson's Foot Cavalry* (New York: Neale Publishing Company, 1912), 44–45.

51. Lomax, *Washington Diary,* 165–70.

52. Betty H. Maury, *The Confederate Diary of Betty Herndon Maury* (Washington, D.C.: privately printed, 1938), 32. Maury's diary is among the rarest and most collectible Confederate books. For details on the book and its author see Robert K. Krick, *The Smoothbore Volley That Doomed the Confederacy: The Death of Stonewall Jackson and Other Chapters on the Army of Northern Virginia* (Baton Rouge: Louisiana State University Press, 2002), 216–22.

53. Joseph A. Brown, *The Memoirs of a Confederate Soldier* (Abingdon, Va.: Forum Press, 1940), 12.

54. Price, *"On to Grafton,"* 56.

55. The Richmond weather source is cited in chapter 1, note 2.

56. Anne S. Frobel, *The Civil War Diary of Anne S. Frobel* (Birmingham: Birmingham Printing & Publishing Co., 1986), 33.

57. Price, *"On to Grafton,"* 57.

58. J. J. Shoemaker, *Shoemaker's Battery* (Memphis: S. C. Toof & Co., n.d.), 12. The bulk of this book, up to the spring of 1864, came from the diary of Lieutenant Edmund H. Moorman, who paid close attention to the weather in his contemporary notes. All citations of *Shoemaker's Battery* in these pages refer to Moorman's primary records.

59. See chapter 1, note 2 for citation of the primary source for Richmond's weather.

60. John B. Beall, *In Barrack and Field* (Nashville: Publishing House of the M. E. Church, South, 1906), 309; Elisha Hunt Rhodes, *All for the Union* (Lincoln, R.I.: Andrew Mowbray, 1985), 46–47.

61. Maury, *Confederate Diary,* 48.

62. L. E. Lee, *Forget-Me-Nots,* 51.

63. Grabill, *Diary,* entries for November 2 and 3; Anderson diary, November 2 and 16, copy in the author's possession.

64. Lomax, *Washington Diary,* 177–78.

65. James V. Drake, *Life of General Robert Hatton* (Nashville: Marshall & Bruce, 1867), 382, 386.

66. Beall, *In Barrack and Field,* 318–19.

67. The report on the meteor appears across the bottom of the first page of the November ledger rather than in the comments or "Observations" section on page 2, but fortunately includes the date preceding the remark.

68. Edmonds, *Journals,* 64.

69. Lomax, *Washington Diary,* 178–81.

70. Drake, *General Robert Hatton,* 389, 395–96.

71. Randolph Barton, *Recollections, 1861–1865* (Baltimore: Press of Thomas & Evans Printing Co., 1913), 18.

## Chapter 3. 1862

The times of sunrise and sunset in the tables for 1862 come from *(Cottom's Edition) Richardson's Almanac, 1862 . . . Calculated by David Richardson, of Louisa County, Va.* (Richmond: J. W. Randolph, [1861]).

1. Walter A. Clark, *Under the Stars and Bars* (Augusta, Ga.: Chronicle Printing Company, 1900), 43; Lucy R. Buck, *Diary of Lucy R. Buck* (n.p., 1940), 7.

2. A copy of the Cumberland weather ledger was filmed for the Asheville archives under the title sheet "U.S. Department of Commerce, Weather Bureau, National Weather Records Center, Asheville, N.C." The reading for January 8 is not legible.

3. Drake, *General Robert Hatton,* 398–99.

4. Clark, *Under the Stars and Bars,* 46.

5. L. E. Lee, *Forget-Me-Nots,* 57.

6. Beall, *In Barrack and Field,* 320–21.

7. Drake, *General Robert Hatton,* 403–7.

8. Clark, *Under the Stars and Bars,* 51.

9. Anderson diary, copy in the author's possession.

10. Benjamin W. Jones, *Under the Stars and Bars: A History of the Surry Light Artillery* (Richmond: Everett Waddey Co., 1909), 23.

11. Gibboney, *Littleton Washington's Journal,* 208.

12. J. B. Jones, *Rebel War Clerk's Diary,* 1:111.

13. The citation for Richmond weather manuscripts is in chapter 1, note 2.

14. Detailed advertisement for sale of "Westwood" in *Richmond Dispatch,* July 9, 1862.

15. Drake, *General Robert Hatton,* 411.

16. White diary in McCarthy, *Richmond Howitzer Battalion,* 2:111.

17. E. V. White, *The First Iron-Clad Naval Engagement in the World* (Portsmouth, Va.: privately printed, 1906). The quotes, all from contemporary witnesses, are scattered through White's slender pamphlet, which consists exclusively of unnumbered leaves.

18. L. E. Lee, *Forget-Me-Nots,* 59–60.

19. Warren, *Eleventh Georgia,* 33.

20. Letter of March 23, 1862, written from Clark's Mountain, in L. E. Lee, *Forget-Me-Nots,* 62.

21. Strider, *George William Peterkin,* 51–52.

22. L. R. Buck, *Diary,* 28–31; Bacot, *A Confederate Nurse,* 97.

23. Diaries of John Waldrop and William Y. Mordecai in McCarthy, *Richmond Howitzer Battalion,* 3:35.

24. The *Wachusett* log is as reported in Edward Powers, *War and the Weather* (Delavan, Wisc.: privately published, 1890), 31. Powers attempted to prove that heavy gunfire brought on rainstorms, and he printed extensive correspondence with Civil War officers (mostly Federals, including such scientific legends as Joshua L. Chamberlain [pp. 152–53]) to make his point. His reporting on manuscript logs kept by Northern warships is of considerable interest and will be cited further below.

25. Warren, *Eleventh Georgia,* 33.

26. Maury, *Confederate Diary,* 65.

27. L. R. Buck, *Diary,* 33.

28. William J. Miller, "Climatological Notes on the Peninsula Campaign, March through August 1862," in *The Peninsula Campaign, Yorktown to the Seven Days,* ed. William J. Miller, vol. 3 (Campbell, Calif.: Savas Woodbury Publishers, 1997), 181. Miller's careful, thorough compilation of a bit of weather information for each day during this period is the best such thing—indeed, the only such thing—done for any extended campaign in the eastern theater. It usually does not supply temperatures, but has other items of interest for most days.

29. George W. Beale, *A Lieutenant of Cavalry in Lee's Army* (Boston: Gorham Press, 1918), 22.

30. White diary in McCarthy, *Richmond Howitzer Battalion,* 2:114.

31. Barbara P. Willis, ed., *The Journal of Jane Howison Beale* (Fredericksburg, Va.: Historic Fredericksburg Foundation, 1979), 38.

32. Richard M. Jolly, *The Story of My Reminiscences* [Gaffney, S.C.: privately printed, 1924], 10–11.

33. James R. Burns, *Battle of Williamsburgh, with Reminiscences of the Campaign* (New York: published by the author, 1865), 18–19, 27, 30; James C. Steele, *Sketches of the Civil War* (Statesville, N.C.: Brady Printing Company, 1921), 20.

34. Waldrop and Mordecai diaries in McCarthy, *Richmond Howitzer Battalion,* 3:38.

35. L. R. Buck, *Diary,* 43–57.

36. Maury, *Confederate Diary,* 78.

37. Samuel D. Buck, *With the Old Confeds* (Baltimore: H. E. Houck & Co., 1925), 32.

38. Miller, "Climatological Notes," 184.

39. Richard Irby, *Historical Sketch of the Nottoway Grays* (Richmond: J. W. Fergusson & Son, 1878), 17–18.

40. William W. Chamberlaine, *Memoirs of the Civil War* (Washington, D.C.: Press of Byron S. Adams, 1912), 16.

41. Joseph P. Thomas, *Memoirs of Joseph P. Thomas* (Richmond: privately printed, [1919]), 4.

42. Eugene A. Nash, *A History of the Forty-fourth Regiment New York Volunteer Infantry* (Chicago: R. R. Donnelley & Sons Company, 1911), 77. Howlett's official service record confirms his death by lightning.

43. Powers, *War and the Weather,* 168.

44. *Harper's Weekly,* July 5, 1862, 426. Bryant is buried in Grave 434 in the Cold Harbor National Cemetery.

45. Maury, *Confederate Diary,* 80–83.

46. For details of the weather at Jackson's two battles, see Robert K. Krick, *Conquering the Valley: Stonewall Jackson at Port Republic* (New York: William Morrow and Company, 1996). The Harrisonburg diary, kept by an unidentified citizen, is little known because of its limited printing, but it is useful for weather purposes, albeit little else. *Harrisonburg, Virginia, Diary of A Citizen, From May 9, 1862–August 22, 1864, Local Events During Civil War* (Harrisonburg, Va.: privately printed, 1961). This listing is from a cover-title, as the book includes no title page.

47. [Beckwith West], *Experience of a Confederate States Prisoner, Being an Ephemeris Regularly Kept by An Officer of the Confederate States Army* (Richmond: West & Johnston, 1862), 10.

48. Powers, *War and the Weather,* 168–69.

49. Lemuel E. Newcomb diary, Newcomb Papers, Stanford University.

50. Fleet and Fuller, *Green Mount*, 139; Willis, *Journal of Jane Howison Beale*, 50.

51. Miller, "Climatological Notes," 187.

52. J. B. Jones, *Rebel War Clerk's Diary*, 1:138.

53. Powers, *War and the Weather*, 35; Miller, "Climatological Notes," 187. The 1 p.m. reading on June 28 also is from the deck log of the *Galena*.

54. David E. Johnston, *Four Years a Soldier* (Princeton, W.Va.: privately printed, 1887), 159.

55. White diary in McCarthy, *Richmond Howitzer Battalion*, 2:124–25.

56. Johnston, *Four Years a Soldier*, 167–68.

57. Powers, *War and the Weather*, 36.

58. Miller, "Climatological Notes," 188.

59. Willis, *Journal of Jane Howison Beale*, 53–55.

60. Powers, *War and the Weather*, 37, 171; Miller, "Climatological Notes," 188.

61. Waldrop and Mordecai diaries in McCarthy, *Richmond Howitzer Battalion*, 3:40–41.

62. Miller, "Climatological Notes," 189. Miller's excellent chapter concludes with a comparison (191) of "Number of Days with Precipitation" in the vicinity on average, for forty years in the mid-twentieth century as opposed to 1862. There were sixty-nine such days during March–August 1862, and sixty-two on average during the long modern sample.

63. For details on the Cedar Mountain weather see Robert K. Krick, *Stonewall Jackson at Cedar Mountain* (Chapel Hill: University of North Carolina Press, 1990). The index (p. 471) cites thirty-two pages that mention the climatic conditions.

64. L. E. Lee, *Forget-Me-Nots*, 68–69.

65. B. J. Haden, *Reminiscences of J. E. B. Stuart's Cavalry* (Charlottesville, Va.: Progress Publishing Co., [1900?]), 15.

66. Mary D. Robertson, ed., *Lucy Breckinridge of Grove Hill: The Journal of a Virginia Girl, 1862–1864* (Kent, Ohio: Kent State University Press, 1979), 28.

67. Shoemaker, *Shoemaker's Battery*, 19.

68. Jedediah Hotchkiss, *Make Me a Map of the Valley: The Civil War Journal of Stonewall Jackson's Topographer* (Dallas: Southern Methodist University Press, 1973), 73–75.

69. W. W. Blackford, *War Years with Jeb Stuart* (New York: Charles Scribner's Sons, 1945), 127.

70. Shoemaker, *Shoemaker's Battery*, 20–21.

71. W. R. Tanner, *Reminiscences of the War Between the States* ([Cowpens, S.C.]: privately printed, 1931), fifth (unnumbered) text leaf.

72. L. E. Lee, *Forget-Me-Nots*, 74.

73. White diary in McCarthy, *Richmond Howitzer Battalion*, 2:130.

74. Powers, *War and the Weather*, 171.

75. Edmonds, *Journals*, 115–16.

76. J. B. Jones, *Rebel War Clerk's Diary,* 1:163.

77. Susan P. Lee, ed., *Memoirs of William Nelson Pendleton, D.D.* (Philadelphia: J. B. Lippincott Company, 1893) 230.

78. Frobel, *Civil War Diary,* 75–84.

79. Fleet and Fuller, *Green Mount,* 171–72.

80. White diary in McCarthy, *Richmond Howitzer Battalion,* 2:136.

81. Fleet and Fuller, *Green Mount,* 175–76.

82. J. B. Jones, *Rebel War Clerk's Diary,* 1:184, 186, 194.

83. White diary in McCarthy, *Richmond Howitzer Battalion,* 2:138; Kimberly Ayn Owen et al., eds., *The War of Confederate Captain Henry T. Owen* (Westminster, Md.: Willow Bend Books, 2004), 99–100.

84. L. E. Lee, *Forget-Me-Nots,* 80.

85. James Dinkins, *Personal Recollections and Experiences in the Confederate Army* (Cincinnati: Robert Clarke Company, 1897), 64–65.

86. E. H. Sutton, *Grand Pa's War Stories* (Demorest, Ga.: Banner Printing Co., [1907]), 17.

87. Wesley Brainerd, *Bridge Building in Wartime* (Knoxville: University of Tennessee Press, 1997), 94–95.

88. Shoemaker, *Shoemaker's Battery,* 24; Cyrus Forwood diary, November 21, 1862, Delaware Public Archives, Dover.

89. Jesse S. McGee to "My dear Mollie," November 24, 1862, copy in the author's possession; Buehring H. Jones, *The Sunny Land* (Baltimore: Innes & Company, 1868), 278–79.

90. William McCarter, *My Life in the Irish Brigade* (Campbell, Calif.: Savas Publishing Company, 1996), 110–11. McCarter served in the 116th Pennsylvania Infantry.

91. J. B. Jones, *Rebel War Clerk's Diary,* 1:204–5, 213, 219–20. A New York infantryman near Fredericksburg made a similar claim that during the night "ice formed thick enough to bear a man." Newton M. Curtis, *From Bull Run to Chancellorsville* (New York: G. P. Putnam's Sons, 1926), 219–20.

92. Forwood diary, December 5, 7, and 15–16, 1862.

93. Oregon D. Foster, 9th Virginia Cavalry, "Reply to Col. Marye," *Fredericksburg Free Lance,* June 7, 1913.

94. Irby, *Nottoway Grays,* 23–24.

95. Herbert Varner of the Macon Light Artillery in the *Macon (Ga.) Telegraph,* January 2, 1863; Oliver S. Coble letter, December 14, 1862, William R. Perkins Library, Duke University. The Georgetown observer timed the appearance of the aurora there at 6:15.

96. Shoemaker, *Shoemaker's Battery,* 26.

97. Elizabeth Maxwell Alsop diary, January 5, 1863, Virginia Historical Society, Richmond. Some of Elizabeth's story, but unfortunately not her diary, is published as

Elizabeth Maxwell Alsop Wynne, *Geneologies and Traditions* (Indiana, Pa.: Printing House of Park, 1931). The misspelling of "genealogies" is in the original title.

98. B. W. Jones, *Under the Stars and Bars*, 66–69.

## Chapter 4. 1863

The times of sunrise and sunset in the tables for 1863 come from *Wynne's Edition, Richardson's Virginia & North Carolina Almanac for the Year of our Lord 1863 . . . Calculated by David Richardson, of Louisa County, Va.* (Richmond: J. W. Randolph, [1862].

1. J. B. Jones, *Rebel War Clerk's Diary*, 1:228, 239–40, 243, 248.

2. B. W. Jones, *Under the Stars and Bars*, 70–71.

3. Hotchkiss, *Make Me a Map*, 109–10; Forwood diary, January 24, 1863; Aaron K. Blake (13th New Hampshire) letter, January 25, 1863, in Civil War Miscellaneous Collection, U.S. Army Military History Institute, Carlisle Barracks, Pennsylvania; Charles S. Wainwright, *A Diary of Battle* (New York: Harcourt, Brace & World, 1962), 159.

4. White diary in McCarthy, *Richmond Howitzer Battalion*, 2:153, 155.

5. J. B. Jones, *Rebel War Clerk's Diary*, 1:255, 258, 262, 266.

6. Shoemaker, *Shoemaker's Battery*, 26–27.

7. Waldrop and Mordecai diary in McCarthy, *Richmond Howitzer Battalion*, 3:46.

8. Strider, *George William Peterkin*, 60.

9. W. H. Ware, *The Battle of Kelley's Ford* (Newport News, Va.: Warwick Printing Co., n.d.), 3.

10. J. B. Jones, *Rebel War Clerk's Diary*, 1:267, 271, 274, 277–78.

11. B. W. Jones, *Under the Stars and Bars*, 85.

12. Edgar Warfield, *A Confederate Soldier's Memoirs* (Richmond: Masonic Home Press, 1936), 142–43.

13. Shoemaker, *Shoemaker's Battery*, 28–29.

14. Rufus H. Peck, *Reminiscences of a Confederate Soldier* [Fincastle, Va.: n.p., 1913], 21–22.

15. *Richmond Sentinel*, March 21, 1863.

16. Warfield, *Confederate Soldier's Memoirs*, 143; Anderson diary, copy in the author's possession.

17. Mary S. Estill, ed., "Diary of a Confederate Congressman, 1862–1863," *Southwestern Historical Quarterly* 39 (July 1935): 39.

18. Robertson, *Lucy Breckinridge*, 107.

19. White diary in McCarthy, *Richmond Howitzer Battalion*, 2:166.

20. Eleanor P. Cross and Charles B. Cross, Jr., eds., *Glencoe Diary: The War-Time Journal of Elizabeth Curtis Wallace* (Chesapeake, Va.: Norfolk County Historical So-

ciety, 1968), 23–25. Unlike many other contemporary records, Elizabeth Wallace's journal pays scrupulous attention to the weather almost every day. Since it only begins with April 1863 (and runs through December 1864), it does not cover the period early in the war when Norfolk and its vicinity stood near the war's center stage. Anyone interested in the weather at the tip of Peninsula during 1863 and 1864, however, should check with Elizabeth.

21. Frobel, *Civil War Diary*, 126–27.

22. L. R. Buck, *Diary*, 146–47.

23. J. B. Jones, *Rebel War Clerk's Diary*, 1:287, 291, 293, 295.

24. Nimrod Newton Nash to "Dearest One," April 5, 1863, copy in the author's possession. Nash served in the 13th Mississippi Infantry.

25. Shoemaker, *Shoemaker's Battery*, 31.

26. Michael Barton, ed., "The End of Oden's War: A Confederate Captain's Diary," *Alabama Historical Quarterly* 43 (Summer 1981): 77.

27. James J. Williamson, *Prison Life in the Old Capitol* (West Orange, N.J.: n.p., 1911), 111–12.

28. Robertson, *Lucy Breckinridge*, 112–13.

29. Hotchkiss, *Make Me a Map*, 137, 139.

30. For a detailed scientific look at the nighttime sky above Jackson's wounding, see Donald W. Olson, "A Fatal Full Moon," *Blue & Gray Magazine* 13, no. 4 (1996): 24–29.

31. White diary in McCarthy, *Richmond Howitzer Battalion*, 2:178–79.

32. Rice C. Bull, *Soldiering* (San Rafael, Calif.: Presidio Press, 1977), 74–75. Bull belonged to the 123rd New York Infantry.

33. J. B. Jones, *Rebel War Clerk's Diary*, 1:321, 326, 331–32.

34. Susan Leigh Blackford, comp., *Memoirs of Life In and Out of the Army in Virginia*, 2 vols. (Lynchburg, Va.: J. P. Bell Company, 1896), 2:42–44.

35. Hotchkiss, *Make Me a Map*, 145–46.

36. Frobel, *Civil War Diary*, 139–40.

37. J. B. Jones, *Rebel War Clerk's Diary*, 1:340, 345, 358, 362.

38. Frobel, *Civil War Diary*, 146.

39. William C. Jordan, *Some Events and Incidents during the Civil War* (Montgomery, Ala.: Paragon Press, 1909), 39.

40. Arthur J. L. Fremantle, *Three Months in the Southern States* (Mobile, Ala.: S. H. Goetzel, 1864), 105, 112, 118–19.

41. Edmonds, *Journals*, 153–54.

42. Charles Mattocks, *"Unspoiled Heart"* (Knoxville: University of Tennessee Press, 1994), 39–40; Joel Molyneux, *Quill of the Wild Goose* (Shippensburg, Pa.: Burd Street Press, 1996), 105–6. Mattocks belonged to the 17th Maine Infantry, Molyneux to the 141st Pennsylvania Infantry.

43. Report from a Philadelphia newspaper, printed in the *Richmond Dispatch*, July 4, 1863.

44. G. W. Beale, *Lieutenant of Cavalry*, 111.

45. The Gettysburg readings come from an excellent, thorough article on weather during the height of this famous campaign: Thomas L. Elmore, "A Meteorological and Astronomical Chronology of the Gettysburg Campaign," *Gettysburg Magazine* 13 (July 1995): 7–21.

46. The Georgetown observations on clouds and wind often proved difficult to decipher, but for this period they are distinct.

47. Elmore, "Meteorological and Astronomical Chronology," 20.

48. Fremantle, *Three Months*, 138–39.

49. G. W. Beale, *Lieutenant of Cavalry*, 111.

50. J. B. Jones, *Rebel War Clerk's Diary*, 1:370–71, 375, 379, 389–90.

51. See note 57 below for citation of the Alexandria records. The count of days by type obviously reached thirty-one between fair and cloudy categories, with the rain days duplicating some portions of the "cloudy" days. In some other months the tabulation considered rainy days as a distinct category, part of the total number of calendar days.

52. Shoemaker, *Shoemaker's Battery*, 49.

53. Hotchkiss, *Make Me a Map*, 162–64.

54. L. E. Lee, *Forget-Me-Nots*, 88–89.

55. J. B. Jones, *Rebel War Clerk's Diary*, 2:4, 9, 27.

56. Waldrop and Mordecai diary in McCarthy, *Richmond Howitzer Battalion*, 3:48.

57. The Alexandria records for the war years cover only from February 1863 to February 1864. The manuscript ledger was filmed in 1952 for the archives in Asheville, under the same title as the Richmond records cited in chapter 1, note 2.

58. Michael B. Chesson and Leslie J. Roberts, eds., *Exile in Richmond: The Confederate Journal of Henri Garidel* (Charlottesville: University of Virginia Press, 2001), 53, 57.

59. B. W. Jones, *Under the Stars and Bars*, 136.

60. L. E. Lee, *Forget-Me-Nots*, 91.

61. Gary Wilson, ed., "The Diary of John S. Tucker: Confederate Soldier from Alabama," *Alabama Historical Quarterly* 42 (Spring 1981): 26.

62. For the Alexandria station's citation, see note 57 above.

63. L. E. Lee, *Forget-Me-Nots*, 93.

64. Shoemaker, *Shoemaker's Battery*, 58–59.

65. White diary in McCarthy, *Richmond Howitzer Battalion*, 2:227.

66. J. B. Jones, *Rebel War Clerk's Diary*, 2:80, 82.

67. L. R. Buck, *Diary*, 191.

68. J. B. Jones, *Rebel War Clerk's Diary*, 2:85, 92–93.

69. Shoemaker, *Shoemaker's Battery*, 61.

70. *Richmond Dispatch,* November 30 and December 3, 1863.

71. Shoemaker, *Shoemaker's Battery*, 63.

72. Wainwright, *A Diary of Battle,* 305.

73. *Richmond Dispatch,* December 3 and 8, 1863.

74. Shoemaker, *Shoemaker's Battery*, 64–65.

75. For the source on the Alexandria station, see note 57 above.

76. J. B. Jones, *Rebel War Clerk's Diary,* 2:119–22.

77. Charles Crosland, *Reminiscences of the Sixties* (Columbia, S.C.: State Company, [1910]), 11–12.

## Chapter 5. 1864

The times of sunrise and sunset in the tables for 1864 come from *The Southern Almanac for 1864, Calculated by David Richardson, of Louisa Co., Va.* (Lynchburg, Va.: Johnson & Schaffter, [1863]). Bibliophiles may be interested to know that the copy used belonged to Charles B. Rouss of the 12th Virginia Cavalry, who later became a New York millionaire and philanthropist.

1. J. B. Jones, *Rebel War Clerk's Diary,* 2:122–25.

2. L. E. Lee, *Forget-Me-Nots,* 103.

3. William W. Cain diary entry, typescript in Bound Volume 321, archives of Fredericksburg and Spotsylvania National Military Park, Fredericksburg, Virginia. Cain served in the 62nd Pennsylvania Infantry.

4. Nathaniel H. Harris, *Movements of the Confederate Army in Virginia and the Part Taken Therein by the Nineteenth Mississippi Regiment from the Diary of Gen. Nat. H. Harris* (Duncansby, Miss.: privately published, 1901), 24.

5. See chapter 4, note 57 for details on the Alexandria station.

6. J. B. Jones, *Rebel War Clerk's Diary,* 2:140–42, 151–52, 161.

7. Shoemaker, *Shoemaker's Battery*, 65.

8. Edmonds, *Journals,* 180, 183.

9. For full citation to the Alexandria source, see chapter 4, note 57. The twenty-nine-day total resulted from 1864 being a leap year. Unfortunately, records for the excellent Alexandria station end here, after running for most of a one-year span.

10. J. B. Jones, *Rebel War Clerk's Diary,* 2:162–63, 165, 167, 178–79.

11. John H. Cammack, *Personal Recollections* (Huntington, W.Va.: Paragon Ptg. & Pub. Co., [1920]), 104.

12. Daniel M. Holt, *A Surgeon's Civil War* (Kent, Ohio: Kent State University Press, 1994), 173–74.

13. Marcus B. Toney, *The Privations of a Private* (Nashville: printed for the author, 1905), 69.

14. Shoemaker, *Shoemaker's Battery*, 69.

15. Waldrop and Mordecai diary in McCarthy, *Richmond Howitzer Battalion,* 3:51.

16. Chesson and Roberts, *Journal of Henri Garidel,* 109–11. A Pennsylvania soldier camped in Culpeper County reported the April 2 storm as "snowing and very bad morning," followed by "cloud & warm" weather on the 3rd. Amos Swart diary, type-script in Bound Volume 198, archives of Fredericksburg and Spotsylvania National Military Park, Fredericksburg, Virginia. Swart served in the 140th Pennsylvania Infantry.

17. The manuscript sheets for "Rendezvous Distribution" were copied as part of the 1952 National Archives job cited for the Richmond records (and several others used in this work). See chapter 1, note 2.

18. J. B. Jones, *Rebel War Clerk's Diary,* 2:188–93.

19. L. E. Lee, *Forget-Me-Nots,* 111.

20. James Conner, *Letters of General James Conner, CSA* (Columbia, S.C.: Presses of the R. L. Bryan Co., 1950), 126.

21. White diary in McCarthy, *Richmond Howitzer Battalion,* 2:240.

22. Chesson and Roberts, *Journal of Henri Garidel,* 129–33.

23. Graves, *Bedford Light Artillery,* 42.

24. Toney, *Privations of a Private,* 78.

25. For a detailed account of the Bloody Angle fighting, including contemporary descriptions of the weather on May 12, see Robert K. Krick, "An Insurmountable Barrier between the Army and Ruin: The Confederate Experience at Spotsylvania's Bloody Angle," in *The Spotsylvania Campaign,* ed. Gary W. Gallagher (Chapel Hill: University of North Carolina Press, 1998), 80–126.

26. Gibboney, *Littleton Washington's Journal,* 265–67.

27. Powers, *War and the Weather,* 60–61.

28. Unidentified Indiana soldier diary, accession no. 13164, Alderman Library, University of Virginia.

29. See note 17 above for the source of the "Rendezvous Distribution" weather manuscripts.

30. *Richmond Whig,* May 12, 1864; J. Staunton Moore, ed. and comp., *Annals of Henrico Parish, Diocese of Virginia, and Especially of St. John's Church* (Richmond: Williams Printing Company, 1904), 49. The church history, which garbles the storm's year, describes the eventual replacement of the steeple in November 1866.

31. J. B. Jones, *Rebel War Clerk's Diary,* 2:207. Jones's entry suggests that he witnessed the violent storm on May 12, but it apparently was the mighty onslaught of late on the 11th, entered on the next day's page, as often happens with diarists.

32. Edward R. Turner, *The New Market Campaign* (Richmond: Whittet & Shepperson, 1912), 85–86.

33. B. W. Jones, *Under the Stars and Bars,* 188.

34. Hotchkiss, *Make Me a Map,* 205–7.

35. Chesson and Roberts, *Journal of Henri Garidel,* 153–61.

36. Jordan, *Some Events and Incidents,* 82–83.

37. Powers, *War and the Weather,* 62.

38. Creed T. Davis diary in McCarthy, *Richmond Howitzer Battalion,* 3:13–14.

39. [A. M. Keiley], *Prisoner of War, or Five Months among the Yankees* (Richmond: West & Johnston, 1865), 5.

40. J. B. Jones, *Rebel War Clerk's Diary,* 2:230–39.

41. Hotchkiss, *Make Me a Map,* 212.

42. Crosland, *Reminiscences of the Sixties,* 20. The June 24 fight also was called Samaria Church, and—by Yankees who could not understand "Samaria" in Southern drawls—St. Mary's Church. Crosland supplied both names.

43. J. B. Jones, *Rebel War Clerk's Diary,* 2:241–51.

44. Diary of Lieutenant William Ashley, Thomas's Legion, in Powers, *War and the Weather,* 165.

45. Davis diary in McCarthy, *Richmond Howitzer Battalion,* 3:16.

46. Chesson and Roberts, *Journal of Henri Garidel,* 188.

47. Hotchkiss, *Make Me a Map,* 218–19.

48. Chesson and Roberts, *Journal of Henri Garidel,* 189–203.

49. J. B. Jones, *Rebel War Clerk's Diary,* 2:262–72.

50. Francis A. Walker, *History of the Second Army Corps in the Army of the Potomac* (New York: Charles Scribner's Sons, 1887), 572.

51. Richard L. T. Beale, *History of the Ninth Virginia Cavalry* (Richmond: B. F. Johnson Publishing Company, 1899), 140.

52. Davis diary in McCarthy, *Richmond Howitzer Battalion,* 3:19–20.

53. G. W. Beale, *Lieutenant of Cavalry,* 181.

54. L. E. Lee, *Forget-Me-Nots,* 121.

55. Chesson and Roberts, *Journal of Henri Garidel,* 205–10, 218–22.

56. Robert E. Park, *Sketch of the Twelfth Alabama Infantry* (Richmond: Wm. Ellis Jones, Book and Job Printer, 1906), 87; Ashley diary in Powers, *War and the Weather,* 166.

57. Letters dated September 11 and 21 in Conner, *Letters,* 151, 154–55.

58. R. Barton, *Recollections,* 63–64.

59. Chesson and Roberts, *Journal of Henri Garidel,* 222, 226–30.

60. J. B. Jones, *Rebel War Clerk's Diary,* 2:297–319. Jones used "bright" repeatedly throughout the month to describe the weather, and "beautiful" nearly as often.

61. Edmonds, *Journals,* 207.

62. Conner, *Letters,* 156.

63. Hotchkiss, *Make Me a Map,* 240.

64. Davis diary in McCarthy, *Richmond Howitzer Battalion,* 3:27; Wilbur Fisk, *Hard Marching Every Day* (Lawrence: University Press of Kansas, 1992), 270.

65. Powers, *War and the Weather,* 63.

66. J. B. Jones, *Rebel War Clerk's Diary,* 2:322–27.

67. Davis diary in McCarthy, *Richmond Howitzer Battalion*, 3:28; Fisk, *Hard Marching*, 273.

68. Chesson and Roberts, *Journal of Henri Garidel*, 232–40.

69. Waldrop and Mordecai diary in McCarthy, *Richmond Howitzer Battalion*, 3:56.

70. [Milton W. Humphreys], *Capt. Thomas A. Bryan, Bryan's Battery* (Richmond: Whittet & Shepperson, n.d.), 23.

71. Chesson and Roberts, *Journal of Henri Garidel*, 244–53.

72. Davis diary in McCarthy, *Richmond Howitzer Battalion*, 3:30.

73. Graves, *Bedford Light Artillery*, 50.

74. Basil L. Neal, *A Son of the American Revolution* (Washington, Ga.: Washington Reporter Print, 1914), 62–64.

75. J. B. Jones, *Rebel War Clerk's Diary*, 2:360–69.

76. *Richmond Dispatch,* January 2, 1865.

## Chapter 6. 1865

The times of sunrise and sunset for 1865 come from *Wynne's Edition, Richardson's Virginia & North Carolina Almanac, for the Year of Our Lord 1865 . . . Calculated by David Richardson, of Louisa County, Va.* (Richmond: West & Johnston, [1864]).

1. Chesson and Roberts, *Journal of Henri Garidel*, 271–77.

2. *Richmond Dispatch,* January 13–14, 1865.

3. J. B. Jones, *Rebel War Clerk's Diary*, 2:383–95.

4. William F. Pendleton, *Confederate Diary of Capt. W. F. Pendleton, January to April 1865* (Bryn Athyn, Pa.: privately published, 1957), 7–8.

5. *Richmond Dispatch,* January 24, 28, 1865.

6. L. E. Lee, *Forget-Me-Nots*, 128.

7. See chapter 3, note 2 for full citation of the Cumberland weather source.

8. Anderson diary, copy in the author's possession; *Richmond Dispatch,* February 11, 1865.

9. J. Staunton Moore, *Reminiscences* (Richmond: O. E. Flanhart Printing Company, 1903), 15, 85. Moore served in the 15th Virginia Infantry.

10. *Richmond Dispatch,* February 8, 15, 1865.

11. J. B. Jones, *Rebel War Clerk's Diary*, 2:422–25.

12. *Richmond Dispatch,* February 24, 27, 1865.

13. Hotchkiss, *Make Me a Map*, 258–60.

14. *Richmond Dispatch,* March 2–4, 8–9, 14, 1865.

15. Chesson and Roberts, *Journal of Henri Garidel*, 334.

16. J. B. Jones, *Rebel War Clerk's Diary*, 2:450–55.

17. R. Barton, *Recollections*, 78.

18. *Richmond Dispatch,* April 1, 1865.

19. Chesson and Roberts, *Journal of Henri Garidel,* 365–79.
20. Graves, *Bedford Light Artillery,* 57, 59, 63.
21. Thomas, *Memoirs,* 19; Harris, *Movements of the Confederate Army,* 44.
22. Davis diary in McCarthy, *Richmond Howitzer Battalion,* 4:4.
23. Moore, *Reminiscences,* 90.
24. Chesson and Roberts, *Journal of Henri Garidel,* 391–98.
25. Davis diary in McCarthy, *Richmond Howitzer Battalion,* 4:10–15.
26. Frobel, *Civil War Diary,* 176.
27. Davis diary in McCarthy, *Richmond Howitzer Battalion,* 4:17–20.
28. Chesson and Roberts, *Journal of Henri Garidel,* 399–404.
29. Edmonds, *Journals,* 226.